Texts in Philosophy
Volume 16

Hao Wang
Logician and Philosopher

Texts in Philosophy Series Editors
Vincent F. Hendriks   vincent@hum.ku.dk
John Symons   jsymons@utep.edu
Dov Gabbay   dov.gabbay@kcl.ac.uk

Hao Wang
Logician and Philosopher

edited by

Charles Parsons

and

Montgomery Link

ISBN 978-1-84890-028-8

College Publications
Scientific Director: Dov Gabbay
Managing Director: Jane Spurr
Department of Computer Science
King's College London, Strand, London WC2R 2LS, UK

http://www.collegepublications.co.uk

Original cover design by orchid creative   www.orchidcreative.co.uk
Printed by Lightning Source, Milton Keynes, UK

Contents

List of Contributors

Martin Davis
(Emeritus, Courant Institute, New
York University; Visiting Scholar,
University of California, Berkeley)
martin@eipye.com

Juliet Floyd
Department of Philosophy
Boston University
745 Commonwealth Avenue
Boston, MA 02215
jfloyd@bu.edu

Marie L. Grossi
272 Silvermine Avenue
Norwalk, CT 06850
mg9711@optonline.net

He Zhaowu
#301 in Unit 1
West-southern Building No. 14
Tsinghua University
Beijing 100084, China

Richard Jandovitz
C. V. Starr East Asian Library
300 Kent Hall
Columbia University
1140 Amsterdam Avenue, MC 3901
New York, NY 10027
rj105@columbia.edu

Eckehart Köhler
Lichtenauergasse 9/6
A-1020 Vienna, Austria
Eckehart.Koehler@univie.ac.at

Montgomery Link
Department of Philosophy
Suffolk University
8 Ashburton Place
Boston, MA 02108
mlink@suffolk.edu

Katalin Makkai
European College of Liberal Arts
Platanenstrasse 24
D-13156 Berlin-Pankow, Germany
khmakkai@gmail.com

Charles Parsons
Department of Philosophy
Emerson Hall
Harvard University
Cambridge, MA 02138
parsons2@fas.harvard.edu

Abner Shimony
(Emeritus, Boston University)
438 Whitney Avenue #13
New Haven, CT 06511
abner.shimony@gmail.com

Preface

The present volume is devoted to the work of Hao Wang (1921-1995), a prolific and wide-ranging worker in mathematical logic, computer science, and philosophy, as well as the person intellectually closest to Kurt Gödel during the latter's later years. Wang's writings did much to document Gödel's thought as well as to interpret it. Since his death rather little had been published on Wang's considerable body of work or on the man's personality and unusual personal history. The present collection could hardly fill this gap but aims to do something toward that end.

A salient fact about Wang is that he was born in China and spent the first twenty-five years of his life there, before coming to the United States for graduate study in 1946. He had by then completed undergraduate and some graduate study; he was sufficiently advanced in logic and philosophy so that he could complete the Ph.D. at Harvard in just two years. He never returned permanently to China, although it appears that at some points he gave serious thought to doing so. Clearly he never lost his identification as Chinese, and we think he resisted being categorized as Chinese-American or Asian-American, although he did become an American citizen in 1967. The influence on him of Chinese culture and thought comes through in a number of his writings, maybe especially in his reaction to the very Western rationalism of Kurt Gödel, which fascinated him but which he could never fully embrace.

One of the aims of the present collection is to present the Chinese side of Wang more directly. Wang had a little publication in Chinese before he came to America, but he did not return to China or publish in Chinese between 1946 and after the opening of relations between the US and the People's Republic, which made it possible for him to travel to China in 1972.[1] He returned to China

---

[1] On this trip see Gödel [2003], p. 388, in the introductory note to the correspondence with Wang. Wang himself wrote about it in print only in Chinese

1

several times after that trip. After 1972 he published a number of essays in Chinese. His writings in Chinese are surveyed in the essay by Jandovitz and Link. Wang's two essays translated here shed particular light on his early life and university studies in wartime China. Jin Yuelin, on whom the earlier essay is centered, was one of Wang's teachers, in particular in logic. The later essay was a preface to a volume of essays by the historian He Zhaowu, one of Wang's closest friends from undergraduate days. We have added a translation of a brief memoir by He written not long after Wang's death, for this is informative as to the character of Wang's intellect and telling as a contrast in styles. One of the themes of these three translations, in evidence also in other Chinese writings that Wang did not choose to publish in English, is the plight of the Chinese intellectual, constrained on the one hand by service to pure research and on the other by service to China and Chinese society.

We have not been able to represent Wang the logician as well as we would have liked. However, the essay by the eminent logician Martin Davis describes two of his classic contributions, first a pioneering result in automatic theorem proving and then his proof that the class of prenex formulae of first-order logic with quantifier prefixes of the form $\forall\exists\forall$ is undecidable. Davis notes that Wang's first involving himself seriously with computers was prompted by thoughts of returning to China. Some other contributions to logic by Wang are briefly noted in the survey essay by Charles Parsons which opens the volume.

Wang's close relationship with Gödel is reflected in two contributions. The first is a memoir by Eckehart Köhler of his collaboration with Wang in the study of Gödel's philosophy. It reveals much about Wang's personality and intellectual style in late years that is not expressed by others who have written about Wang. The second, edited with an introduction by Charles Parsons, is a previously unpublished text by Wang undertaking to summarize views of Gödel on sets and concepts. It was written in late 1975, during Wang's second series of conversations with Gödel, and the handwritten emendations (some of them by Gödel himself) reflect Gödel's reactions. The introduction gives information about other "fragments" Wang wrote at the time, which were either incorporated

---

(see [1972]), and his attitude toward the regime of the People's Republic was more favorable then than it was several years later.

into other writings (including this one) or are of less interest.

The book closes with two essays on Wang's philosophy. Abner Shimony presents the most extended discussion of Wang's conception of philosophy and philosophical method known to us. The final essay, by Juliet Floyd, deals with Wang's long engagement with the philosophy of Wittgenstein. This helped him to articulate his vision of the development of knowledge in terms of a dialectic between the intuitive and the formal, practice and theory, logic and philosophy. He also contributed distinctive and insightful interpretations of Wittgenstein's thought. Floyd concludes with reflections on Wang as a teacher.

In preparing this book we have incurred many debts. The primary one is to Wang's widow Hanne Tierney-Wang, who has encouraged our interest in Hao Wang's legacy and who has graciously permitted the publication of writings of Wang either previously unpublished or not previously published in English. We are also indebted to Zhu Wenman for assistance and advice concerning the Chinese side of our publication.

Our project originated with the proposal of Jaakko Hintikka, then editor-in-chief of *Synthese*, of an issue of that journal devoted to Wang's work. The project could not be realized at that time. We are grateful to the present editors for allowing its revival and to one in particular, John Symons, for helpful correspondence and especially for suggesting the publication of our collection as a book and for smoothing the way to its appearing with College Publications. We are also grateful to Marc Gasser for his expert typesetting.

Jane Spurr, Managing Director of College Publications, has been unfailingly helpful in matters pertaining to the publication arrangement and the production of the book that is now in your hands.

Hao Wang was not honored with a Festschrift during his lifetime. Had such a volume been put together, it would have been of a quite different nature from the present one. Nonetheless, although only one of us knew him personally,[2] we would like to think of this

---

[2](*Note by Charles Parsons.*) I was Wang's student during one academic year, 1955-56. About that I have written the following:

> In the fall of 1955, as a first-year graduate student at Harvard, I took a seminar with him on the foundations of mathematics. I had begun the previous spring to study intuitionism, but without

4

volume as a tribute to a remarkable personality, with many-sided contributions to logic, computer science, and philosophy.

Charles Parsons
Montgomery Link

---

much context in foundational research. Wang supplied some of the context. In particular he lectured on the consistency proof of Ackermann [1940]. I knew of the existence of some of Kreisel's work (at least Kreisel [1951]) but was, before Wang's instruction, unequipped to understand it. In the spring semester, in a reading course, he guided me through Kleene [1952]. Unfortunately for me he left Harvard at the end of that semester, but his instruction was decisive in guiding me toward proof theory and giving me a sense of its importance. (Parsons [1998], note 10)

Near the end of that year he introduced me to Paul Bernays. His instruction did not quite end with his departure; we had some correspondence, and one letter written the following year made a suggestion that was very important in setting the direction of my dissertation.

Wang was again at Harvard during my time in residence as assistant professor (1962-64). He participated in the "logic lunches" in those years, where the conversation ranged widely over contemporary work in logic as well as the history of logic. (Harvard logicians, particularly Burton Dreben, played a prominent role in the planning of Jean van Heijenoort's *From Frege to Gödel* [1967], and van Heijenoort himself was present at the lunches from time to time.) Not long after my move to Columbia University, Wang also moved to New York, to Rockefeller University. We were by then friends, and our conversations covered a wide range of topics in logic and philosophy.

References

Ackermann, Wilhelm, 1940. Zur Widerspruchsfreiheit der Zahlentheorie. *Mathematische Annalen* 117, 162-194.

Gödel, Kurt, 2003. *Collected Works*, Volume V. *Correspondence H-Z*. Edited by Solomon Feferman, John W. Dawson, Jr., Warren Goldfarb, Charles Parsons, and Wilfried Sieg. Oxford: Clarendon Press.

Kleene, S. C., 1952. *Introduction to Metamathematics*. New York: Van Nostrand.

Kreisel, G., 1951. On the interpretation of non-finitist proofs, part I. *The Journal of Symbolic Logic* 16 (1951), 241-267.

Parsons, Charles, 1998. Hao Wang as philosopher and interpreter of Gödel. *Philosophia Mathematica* (3) 6, 3-24.

Van Heijenoort, Jean, 1967. *From Frege to Gödel: A Source Book in Mathematical Logic*, 1879-1931. Cambridge, Mass.: Harvard University Press.

Wang, Hao, 1972. Reflections on a visit to China (Chinese). *New China Bi-monthly*, no. 7 (October 1), 23-26 and 31.[3]

---

[3]For information on revisions and reprintings see the Bibliography in this volume.

Hao Wang[1]

Charles Parsons

Hao Wang (1921-1995) is known for his contributions to mathematical logic, computer science, and philosophy. He was a native of China and came from there to the United States in 1946. Except for a five-year interval in England, he remained in the US for the remainder of his life. After the opening up of relations between the US and the People's Republic of China, however, he renewed his own relations with China and visited there already in 1972, and a number of times thereafter. Although he became a US citizen in 1967, Wang would have resisted characterization as an Asian-American. I believe he thought of himself as simply Chinese, a member of the Chinese diaspora that has existed for centuries.

Wang was born in Jinan, Shandong, China, May 20, 1921. He obtained a B.Sc. in mathematics and an M.A. in philosophy in wartime China.[2] In 1946 he came to Harvard to study logic and philosophy. He received his Ph.D. in 1948 and was a Junior Fellow of the Society of Fellows at Harvard until 1951. From then until 1961 he taught philosophy at Harvard and then Oxford. He returned to Harvard in 1961 as Gordon McKay Professor of Mathematical Logic and Applied Mathematics. But in 1966 he went to the Rockefeller University as a visiting professor; the next year he became professor, establishing a research group in logic. He made Rockefeller an active

---

[1]Reprinted with slight changes from the newsletter *The Status of Asian/Asian-American Philosophers and Philosophies*, American Philosophical Association, vol. 01, no. 2, 2002, by permission of the American Philosophical Association. The previous title was "Hao Wang and mathematical logic."

[2]Wang writes a little about his studies in China in [1982b] and [1993f], of which translations appear in this volume. In matters concerning Wang's Chinese writings, I am greatly indebted to Montgomery Link. See now the essay by Richard Jandovitz and Link, "Hao Wang's Chinese writings," in this volume.

center, especially of research in set theory. After the group was broken up by the Rockefeller administration in 1976, only Wang remained, even beyond his retirement in 1991. He died in New York May 13, 1995.

Wang was a philosopher from early on and published his first philosophical essay before he left China. However, the primary field of his early work was logic, and his publications through the early 1960s are largely in mathematical logic. He published a large number of papers, most of which up to 1960 are included in *A Survey of Mathematical Logic* [1962a]. One significant contribution arose from W. V. Quine's attempt in his book [1940] to add classes to the sets of his well-known system New Foundations (NF). The axiom Quine proposed was shown inconsistent by J. Barkley Rosser in 1942. Wang analyzed the situation thoroughly and devised the axiom that best expressed the intended idea, which was then incorporated into the revised edition [1951]. Wang gave a model-theoretic proof that if NF is consistent then his revision is also consistent.

Perhaps encouraged by the year (1950-51) that he spent in Zürich under the auspices of Paul Bernays, Wang worked throughout the 1950s on questions of the relative strength of axiom systems, particularly set theories. He was a pioneer in the post-war research reviving Hermann Weyl's idea that mathematics might be developed in a way that avoids impredicative set existence assumptions. He also contributed to the effort of logicians of the time to analyze predicative definability.

Wang gained practical experience with computers early on, and some of the papers he published around 1960 are significant work on the border between logic and computer science, long before "logic in computer science" became a field with hundreds of publications every year. The best known of these papers, [1960a], reports programs that proved all the theorems of propositional and predicate logic in *Principia Mathematica* in a few minutes. By using the kind of logical analysis pioneered by Herbrand and Gentzen, he was able to improve substantially on the previous work of Newell, Shaw, and Simon. Possibly his most significant result in mathematical logic was the proof, obtained with A. S. Kahr and E. F. Moore in 1961 (see [1962b]), that the general decision problem for first-order logic can be reduced to that for the class of quantificational formulas of the form "For all x, some y, and all z $M(x, y, z)$," where M contains no quantifiers, so that satisfiability of formulas in that class is undecidable.[3]

Wang's prolific writing in logic included expository and historical work, which is to be found in [1962a] and in some of his philosophical writings, especially *From Mathematics to Philosophy* ([1974a], hereafter FMP). But he wrote only one expository book on logic, [1981a], based on lectures given in China.

Wang's early philosophical writings are short critical pieces, varied in content.[4] Longer pieces in the 1950s stay close to logic and the foundations of mathematics but express a point of view owing much to the European work before the Second World War. Probably his first really distinctive extended philosophical essay is "Process and existence in mathematics" [1961b]. This essay clearly reflects reading of Wittgenstein's *Remarks on the Foundations of Mathematics*, although Wittgenstein's name is not mentioned. The notion of perspicuous proof, the question whether a mathematical statement changes its meaning when a proof of it is found, the question whether contradictions in a formalization are a serious matter for mathematical practice and applications, and a Wittgensteinian line of criticism of logicist reductions of statements about numbers are all to be found in Wang's essay. But it could only have been written by a logician familiar with computers. Computers and Wittgenstein enable Wang to present issues about logic in a more concrete way than is typical in logical literature then or later.

This essay also exhibits a style characteristic of Wang's philosophical writing, which is to present a certain amount of the relevant logic and mathematics, to look at the issues from different angles and to discuss different views of the matter, very often without pressing very hard an argument for a particular view, even keeping a certain distance from the points of view presented. He was clearly skeptical of systematic theory in philosophy, such as was aspired to by major philosophers of the past and by his teacher W. V. Quine and his friend Michael Dummett. He also resisted the idea, prevalent in earlier post-war analytic philosophy, of "solving philosophical problems" by local analyses and arguments. But he did think that "philosophy should try to achieve some reasonable overview" (FMP, p. x). One could say he aspired to be synoptic

---

[3]On these matters see the paper by Martin Davis in this volume.

[4]Most are reprinted as an appendix to FMP.

but not to be systematic.

Wang's most important statement in the philosophy of logic and mathematics is FMP, which was probably begun not long after his move to Rockefeller. It was also during that period that Wang's extended series of exchanges with Kurt Gödel began, and these are reflected importantly in the book. But much of the book does not reflect Gödel's influence at all. Wang describes his point of view as "substantial factualism," the thesis that philosophy should respect existing knowledge. Although mathematical and scientific knowledge have special importance, they are not uniquely privileged as in some naturalistic views, and the autonomy of mathematics is maintained. Wang's method is also much more descriptive than that envisaged, for example, in Quine's program of naturalistic epistemology. In the context of FMP, this led Wang to include quite a bit of exposition of the relevant mathematical logic, as well as some of its history.

The book covers a number of basic topics in and related to the foundations of mathematics: the concept of computability, logical truth, the concept of set, the mathematical importance of computers, the mechanistic theory of mind. Probably the most notable chapter is that on the concept of set, where Wang explores the conceptual foundations of the accepted axioms of set theory more thoroughly than others had before in print. The chapter on minds and machines expresses a skeptical view of the claims both of mechanists and anti-mechanists and reports some views of Gödel on the subject. Wang explored these themes more fully in the late essay [1993d]. Probably his main contribution was to make clear how many difficult questions surround the "machine" side of the proposed equation of mind and machine.

Taken as a whole, FMP offers a case, based on the general idea of factualism, for rejecting the pictures of mathematics offered by the logical positivists, by Quine, and by many lesser philosophers. It does not proceed by arguing directly with their views, but rather by laying out some of the data that in Wang's view they have failed to consider adequately.

Wang conducted an extended series of conversations with Gödel during the time that he was finishing FMP. Gödel's influence is in evidence in some important places, particularly in the chapter on the concept of set and the chapter on minds and machines. But

Wang also presents some previously undocumented views of Gödel. The introduction quotes most of two letters of Gödel written in 1967 and 1968. Views of Gödel are presented there, in the discussion of computability, and in parts of the chapter on the concept of set. The chapter on minds and machines contains a section, "Gödel on minds and machines." In each case the presentation of Gödel's views was approved by him (see p. x). Not only were some of these passages revised by Wang in response to Gödel's comments, but in some cases the final drafts were sent to the publisher by Gödel; in particular that was true of the above-mentioned section.[5] This unusual arrangement reflected an unusual relationship, the closest intellectual relationship into which Gödel entered in his later years.

Wang's relation to Gödel influenced all his subsequent philosophical work, which is reflected in three further books. The first, *Beyond Analytic Philosophy* [1985a], is a structured around a meditation on the philosophical development from Bertrand Russell through Rudolf Carnap to Quine. Wang thought it symptomatic of a general philosophical climate that the interests manifested in Russell's body of writing were so much broader than those of Carnap and Quine and nearly all later analytic philosophers. He thus sees the later development as a history of "contraction." On the other hand he maintains (as have many others) that Russell did not succeed in his later career in working out a philosophical project of the depth and impact of his early work centered on logic. The book is a rich discussion of aspects of the development of twentieth-century philosophy. The treatment of Quine could be objected to as too unsympathetic and as not engaging Quine's arguments. But the introduction articulates and criticizes in an interesting way, influenced by Gödel, assumptions common to Carnap's and Quine's treatment of analyticity. As in other writings, the last part of the book meditates on the nature of philosophy.[6]

Wang viewed *Beyond Analytic Philosophy* as the first of a trilogy, of which the second was to be called *Reflections on Kurt Gödel*, and the third was to deal with Gödel and Wittgenstein as exempli-

---

[5]See Wang's letters to Gödel of April 10 and 26, 1972, and Gödel's letters to Ted Honderich of June 27 and July 19, 1972, as well as §2.3 of the introductory note to the Wang correspondence, in Gödel [2003].

[6]On Wang's philosophical method see also §1 of Parsons [1998], and the essay by Abner Shimony in this volume.

12

fying a "quest for purity" in philosophy. Wang would in these three books have discussed extensively all the figures of twentieth-century philosophy who engaged him most deeply. The second book appeared as [1987a]. It was the first real attempt to see Gödel whole, as a logician, mathematician, philosopher, and to some extent anyway as a man, reflecting his conversations with Gödel as well as access to some then unpublished documents. It has probably been the most widely read of Wang's books.

The remainder of Wang's life was largely taken up by another project concerning Gödel, a presentation and discussion of Gödel's philosophical views using his record of their conversations. Wang reconstructed from his notes a number of remarks from the conversations, which give information about Gödel's philosophy and view of the world that are not otherwise documented. This alone makes the resulting book, *A Logical Journey: From Gödel to Philosophy* [1996a], published after Wang's death, an important work. But the accompanying commentary is not only of great value as interpretation but expresses the response of a philosophical mind that was more broadly trained and informed than Gödel's own and, although very respectful of Gödel and his rationalistic position, had really a very different conception of philosophy, as I have said less systematic and more descriptive.

It is to be regretted that Wang did not live to write the third work in his trilogy.[7] In his later thinking, he was fascinated by Gödel's rationalism but always fell short of embracing it, and he continued to think about Wittgenstein's very opposed philosophy, skeptical of any such systematic construction as Gödel aspired to.

After his first return to China in 1972, Wang began to publish again in Chinese. Evidently what he published were largely short essays, and much of what did not duplicate his writings in English consisted of essays on Chinese intellectual figures, in most cases his own teachers and friends. In many of Wang's philosophical writings, one can sense his Chinese cultural background, but even his explicit statements about Chinese thought are best commented on by someone familiar with the Chinese language and intellectual tradition. It is to be hoped that some scholar with the relevant training will take up this task.

---

[7]The paper [1991b] may have come from this project.

References

Cited writings of Hao Wang:[8]

1960a. Toward mechanical mathematics. *IBM Journal of Research and Development* 4, 2-22.

1961b. Process and existence in mathematics. In Y. Bar-Hillel, E. I. J. Poznanski, M. O. Rabin, and A. Robinson (eds.), *Essays on the Foundations of Mathematics*, dedicated to Prof. A. A. Fraenkel on his 70th anniversary, pp. 328-351. Jerusalem: Magnes Press, The Hebrew University of Jerusalem. Partly incorporated into FMP, ch. 7.

1962a. *A Survey of Mathematical Logic*. Peking: Science Press, 1962. Amsterdam: North-Holland, 1963.

1962b. (With A. S. Kahr and Edward F. Moore.) Entscheidungs-problem reduced to the $\forall\exists\forall$ case. *Proceedings of the National Academy of Sciences U. S. A.* 48, 365-377.

1974a. *From Mathematics to Philosophy*. London: Routledge & Kegan Paul. Cited as FMP.

1981a. *Popular Lectures on Mathematical Logic*. Beijing: Science Press, and New York: Van Nostrand Reinhold. (A Chinese translation was published almost simultaneously.[9])

1982b. Memories related to Professor Jin Yuelin (Chinese). *Wide Angle Monthly*, no. 122, 61-63. English translation in this volume.

1985a. *Beyond Analytic Philosophy. Doing Justice to What We Know*. Cambridge, Mass.: MIT Press.

1987a. *Reflections on Kurt Gödel*. Cambridge, Mass.: MIT Press.

1991b. To and from philosophy—Discussions with Gödel and Witt-genstein. *Synthese* 88 (1991), 229-277.

1993d. On physicalism and algorithmism: Can machines think? *Philosophia Mathematica* (3) 1, 97-138.

---

[8]For a full bibliography of Wang's writings, see the bibliography in this volume.

[9]The statement of Parsons [1996], p. 110, that this work was first published in Chinese is incorrect.

1993f. From Kunming to New York (Chinese). *Dushu Monthly* (May), 140-143. English translation in this volume.

1996a. *A Logical Journey. From Gödel to Philosophy.* Cambridge, Mass.: MIT Press.

Other cited writings:

Gödel, Kurt, 2003. *Collected Works*, volume V: *Correspondence H-Z*. John W. Dawson, Jr., Solomon Feferman, et al., eds. Oxford: Clarendon Press.

Parsons, Charles, 1996. In memoriam: Hao Wang, 1921-1995. *The Bulletin of Symbolic Logic* 2, 108-111.

Parsons, Charles, 1998. Hao Wang as philosopher and interpreter of Gödel. *Philosophia Mathematica* (3) 6, 3-24.

Quine, W. V., 1940. *Mathematical Logic.* New York: W. W. Norton.

Quine, W. V., 1951. Revised edition of [1940]. Cambridge, Mass.: Harvard University Press.

Hao Wang's Chinese Writings[1]

Richard Jandovitz and Montgomery Link

Of the 174 items in the bibliography of Wang's writings, twenty-five are publications in Chinese.[2] However, [1976a] had been originally published in English, and [1987f] and [1987g] were translated from English by another. Furthermore, [1973a] is of dubious authorship.[3] The remaining items fall into four categories:

1. Juvenilia, essays written before Wang's departure in 1946 to study at Harvard University.

2. Reminiscences, essays on his teachers and classmates from his student days in China, either as introductory essays to one of their newly published books or as conference talks celebrating their milestones, or writings expounding the views of Lu Xun.[4]

3. Two writings upon his return to China after an absence of twenty-six years, [1972] and [1979c].

---

[1]Thanks to Charles Parsons, and to Tu Weiming for the connection between Wang and Lu Xun.

[2]We refer to the bibliography in this volume.

[3]It is contained in the initial bibliography upon which Grossi et al. [1998] is based (as item number 79), yet the author of the work, about whom we have no further information, is Xu Shangwen, not Wang, and the article discusses life in Hong Kong, while Wang, although he did visit at least once, had little expertise there, so we mention it no further in this introduction. The remaining Chinese writings are Wang [1944], [1945], [1972], [1977b], [1977c], [1979b] (translated in [1990a]), [1979c], [1981b], [1981e], [1981f], [1982a], [1982b], [1986a], [1987c], [1987d], [1987e], [1989b], [1990e], [1993b], [1993c], and [1993f]. We have not been able to locate for verification [1977c]. It may be that there are other Chinese works by Wang not listed in the bibliography, but evidence to that effect seems to have gone missing.

[4]Including: [1982b] and [1987e] on Jin; [1977b], [1982a], [1989b], and [1990e] on Lu; plus [1981b], [1987c], for which see [1993g], and [1993f].

15

    4. Writings on purely philosophical or technical subjects, all of
       which he had written about in English.

The writings in category (2) are those of greatest interest, but we
will give information about the other writings.

    In all of his Chinese writings, across all categories, Wang returns
again and again to the theme of the dilemma facing the Chinese intellectual since the Opium War in 1840. The style in his Chinese
writings will be recognizable to those who are familiar with his
writings in English: a proliferation of ideas, sometimes complementary, often contradictory, presented in such a manner as to obscure
the point where those ideas leave off and Wang's own ideas begin.
Often Wang will present his ideas with the disclaimer that he has
not thought them through clearly yet and is merely attempting to
stimulate debate. The debate he is most interested in stimulating
in his Chinese writings is how best to modernize China, especially
in its treatment of intellectuals. However, several of these writings
are also informative about Wang's early life.

    Offering some biographical detail on his own life in "Memories
related to Professor Jin Yuelin" [1982b], translated here, Wang relates how his father urged him in high school to read books by the
likes of Feuerbach, books Wang found incomprehensible. When in
1938 he picked up Jin's book *Logic*, though, he found it very easy,
particularly the chapter on mathematical logic.[5] He decides, he
says, "to study the easy first in preparation for the harder later."
Wang's old friend He Zhaowu notes in "Remembering Wang Hao"
the material hardship that Wang endured as a student in China.[6]
Wang attended Xinan Lian, Southwest Associated University, in
Kunming from 1939 to 1946. The University was a wartime association of Peking University, Tsinghua University, and Nankai
University. Wang's "virgin work," as he calls it, was an unpublished paper in Chinese on "Hume's problem of induction," written
in 1942 or 1943 while an undergraduate, now long lost. Also lost is
his Master's Thesis, finished by the spring of 1945, also in Chinese,
which he says in that same article was more four separate essays
than one coherent whole, a scattered style he laments he found
difficult to shake.

---

[5] Jin [1937].
[6] He [1995], translated in this volume.

The first publication listed in the Wang bibliography is "The metaphysical system of the New Lixue" [1944].[7] The Lixue, "school of principle," was a Neo-Confucian school of the Sung and Ming dynasties. The comprehensive system of the New Lixue results from Fung Yu-lan's reconstruction of rational Neo-Confucianism. Fung, a member of the faculty of Tsinghua University and the preeminent philosopher in China of the twentieth century, was one of Wang's teachers and friends.[8]

In his second publication Wang examines the connection between "Language and metaphysics" [1945].[9] He argues against the position he attributes to Russell's *An Inquiry into Meaning and Truth* that an empiricist study of the structure of language can lead to metaphysical knowledge.[10] This is the earliest account of Wang's assessment of the failure of the modern empiricist attempt to use sense data as a guide to scientific inquiry. There follows a gap of a quarter-century in which Wang does not publish in Chinese at all. We now turn to the writings in category (2).

In "The searchings of Lu Xun" [1977b], Wang writes about the most important literary and cultural figure of twentieth-century China, holding him up as a model for all intellectuals to follow, having allowed no bounds to constrain him within narrow definitions of genre or profession. He finds in Lu Xun a hard-headed refusal to knuckle under, a spirit unafraid of sacrifice, a resolute tenacity, all in service of making a contribution to the cause of saving China. Wang provides capsule summaries of Lu's writings and personal experiences, gleaning life lessons on such topics as revolutionary love (loving not a single individual or group but the entire masses), the ability of the idealist who holds on to his principles to persevere in the face of despair, the necessity to focus on concrete problems and not simply apply any one theory wholesale, and the importance of studying history so as to apply historical precedent to current situations.

In "Remeeting Mr. Shen Congwen" [1981b], Wang writes of his

---

[7]Note that the bibliographical entry in Grossi et al. [1998] has been adjusted.

[8]For further information on Fung see "From Kunming to New York," n. 5, in this volume.

[9]Now translated as [2005]. Wang will return to Russell with more philosophical experience in two English articles, [1965a] and [1966d], as well as in [1985a].

[10]Russell [1940].

old acquaintance from university days, who had just given a talk at Columbia University. Wang praises Shen for producing, after 1949, excellent work in fashion history and cultural artifacts, a field completely different from that of his literary work before 1949. Wang sees in this accomplishment proof that one can succeed in different fields in a single lifetime, even if doing two things at once is probably too difficult.

This brings Wang to mention Jin Yuelin, regretting that he had not continued the work he had been doing before 1949, ruing the decline in "pure" learning of the kind Wang thinks China needs now. He recalls his dislike of the phrase Shen had used in those days about Jin, "spiritual nobility," a phrase Wang now sees in a different light, thinking of Germany at the turn of the nineteenth century, a poor and weak country yet able to produce such nobles of the spirit as Goethe, Kant, Beethoven, Schiller, Hegel, and so forth. China, he concludes, is full of intelligent, hard-working people; the problem is how to remove the obstacles preventing them from developing their talent.

Wang returns to Lu Xun, commemorating the centennial of his birth in "To confirm some impressions by Lu Xun" [1982a]. Warning against turning historical figures into paragons of virtue, as propagandists are wont to do, lest the youth find such a standard too hard to live up to and quit trying, Wang notes Lu's contradictions and failures. Critical of others and of China's backwardness, Lu nevertheless held steadfast to his belief that once the patient, a sick China, was stabilized it would produce such giants as Newton, Shakespeare, Darwin, and so forth. Wang once again writes that he himself thinks China has undoubtedly the world's most abundant human resources and human potential, as evidenced by the excellence of the graduate students in all the West's best universities. He believes that if China reduces its reliance on rote learning and encourages its youth to develop their individual interests and talents, it will not only help produce such luminaries but also serve China's long-term interests.

In his reminiscence of Jin Yuelin, previously mentioned, Wang writes of his mentor, Professor Jin, on the occasion of his 55th year with the Institute of Philosophy at the Chinese Academy of Social Sciences. In 1958 Jin visited England and addressed the philosophy faculty, telling them that since Marxism had saved China he had abandoned the philosophical work he had been doing and become a Marxist. The faculty thought his talk a little simplistic, but when Wang pressed him to explain Marxism in more detail Jin replied to the effect that Marxism was not the kind of thought that can be explained better than it can be thought of in one's head.

In "The Way of Jin Yuelin" [1987e], Wang returns once more to his teacher, comparing the two competing (and, Wang says, incompatible) ideals Jin worked on before and after 1949. Before 1949 the ideal was to use philosophy as a specialized area of research that could directly and indirectly raise China's standing in international philosophical circles; after 1949 the ideal was to use philosophy as a kind of "weaponized" thought to directly serve the immediate needs of the nation. Especially in philosophical investigations, Wang writes, relatively fundamental research and the needs of current propaganda are often far apart, likening the post-1949 work of philosophers like Jin to a high-level mathematician or physicist asked to tinker in a garage to churn out a saleable product.

"A reading of Wang You-qin on Lu Xun" [1989b] and "Between philosophy and literature" [1990e] take up once again Wang's interest in Lu Xun, stimulated in part by the publication of Wang Youqin's monograph examining his work in connection to the question of China's national characteristics. Wang looks at the three big questions Lu was said to have focused on: what is the ideal human nature, what is the biggest flaw in the Chinese national character, where is the root of the sickness? Wang compares Lu to the kind of philosopher who shows rather than says, finding in his works a comprehensive picture of the problem, laying out how the characteristics in question operate beneath the level of consciousness, an awareness of which expands the scope of vision and enlarges the field of choice.

"From Kunming to New York" [1993f], translated here, is a reminiscence of He Zhaowu, Wang's classmate born in 1921 and still associated with the Department of History at Tsinghua University. Wang mentions He's plan to "write several systematic books", a plan He continues to realize with publications in the last few years. "From Kunming to New York" touches on many of the themes that dominate Wang's writings in Chinese: the plight of the Chinese intellectual and Wang's abiding concern for genuine academic discourse in China; his infatuation with Marxism, believing it the savior of China, and later disillusionment with it; his own ambitious wish, and admission of failure, to combine the insights of his specialized work with a more general and comprehensive philosophy of life; and his respect for literature in presenting just such a comprehensive philosophy. We next turn to the writings in category (3).

In "Reflections on a visit to China" [1972], after a 1972 return to China, his first after leaving for Harvard in 1946, Wang writes about the changes in his homeland.[11] Acknowledging the limits of his ability to understand Marxism and the limits on his ability to experience life under it more freely than he could in an organized four-week tour, he nevertheless credits the thought of Marxism and the actions of the Chinese Communist Party for succeeding where generations of Chinese intellectuals since the 1840 Opium War had failed. He attributes the successes partly to the close fit between Marxism and traditional Chinese thought, especially the notion of *qusi ligong* (eliminate the private, establish the public) in order to realize the ideal of *tian xia wei gong* (all the world is a commonwealth).

In Beijing Wang reunited with many classmates of his era and marvels at the amount of talent pre-1949 China produced, worrying that the output and use of human resources after 1949 had lagged behind. Part of the problem here, he writes, is the stress on *lilun peihe shiji* (theory in concert with practice), reducing the number of intellectuals doing pure theoretical work. He finds in a few comments from friends hope that that practice is changing and in the future China will be stressing theory a little more.

In "China today and its development over the last sixty years" [1979c], Wang takes the occasion of the sixtieth anniversary of the May 4th Movement of 1919 to reflect on the intervening years and the problems that still face China. Under the new slogan "Seek Truth from Facts" he recognizes the need to examine the current circumstances truthfully in order to find the suitable concrete measures needed to change Chinese society for the better. A recurring theme throughout is the possibility of finding a point, or points, of wrong turning on the road to development, such as those that led to the anti-rightist campaign and Great Leap Forward of the late 1950s, and taking another path. These events may have contributed

---

[11]Cf. [1974c].

to Wang's abandoning the ideas of return he evidently considered in the early to mid-1950s. We now conclude with the writings in category (4).

In [1976a] Wang offers a "recipe" for designing and building a high-speed Chinese typing machine, but this, as mentioned at the outset, was published previously in English in [1990a] as [1973b]. In [1979b] he produces a "Mechanical treatment of Chinese characters" that appeared later in English as "On information processing in the Chinese language." "Gödel and Wittgenstein" [1981e] covers much the same ground as the English talk [1987b] and elsewhere, although in somewhat less detail. Wang notes that this lack of detail is unfortunate, particularly concerning the impression he might have left of certain weaknesses in Wittgenstein's philosophy, saying he has treated but a weak sliver of the subject and that not comprehensively at all and is preparing to write more on the subject. A translation of this work is in progress. In "Mathematical logic" [1981f] Wang discerns four branches of mathematical logic and gives a brief historical overview. He begins with Frege's *Begriffsschrift* but mainly follows the advance in set theory on the continuum hypothesis from Cantor through Gödel to Cohen.[12] There is philosophical discussion that is expanded on in "Philosophy through mathematics and logic" [1990b].

"Chinese and western philosophy" [1986a] is in two parts, the second of which is a translation of "Two commandments of analytic empiricism" [1985b]. Wang divides part one into four sections: 1) philosophy and human life, asking what philosophy does and comparing it to religion, literature, and history; 2) philosophy and society, beginning with Kant's advocacy of individual autonomy and proceeding to connect science with "modernization" before concluding with the idea that the very establishment of such ideals as equality, democracy, tolerance, and so forth, not only gives humanity ends to aspire toward but also the means to attain those ends; 3) philosophy and knowledge, where he writes that the biggest difference between Chinese and Western philosophy is that the latter emphasizes systematic theoretical knowledge more than the former, which emphasis goes a long way toward explaining the appearance of theoretical science in Europe as a unique historical event; and 4)

---

[12]Frege [1879].

Chinese and Western culture, in which he questions just which part of Chinese culture has hindered its modernization before summarizing the differences between the two: China emphasizes human life, direct experience of aesthetic beauty, synthesis, and harmony, while the West emphasizes knowledge, systematization of theoretical ideas, analysis, competition. He concludes with the thought that the real difference between East and West is that of ancient and modern. Further insights in English are in "Philosophy: Chinese and western" [1983] and "Thought and action" [1984d].

In "On distinguishing problems of different orders" [1987c], Wang, addressing an unspecified conference, from the context most likely a group of overseas Chinese intellectuals, writes specifically to discuss Chinese culture by examining the relationship between the traditional and the modern. This is the 100-year topic: how to integrate China and the West. Leaving off concerns about cultural content and questions of relative value to ask more "second order" questions, Wang suggests a fundamental difference between China and the West is the former's emphasis on synthesis (and harmony and continuity) versus the latter's emphasis on analysis (and contradiction and disruption). The Chinese work [1987d] on Einstein and Gödel covers the same ground as the English "Gödel and Einstein as companions" [1991d]. Much of the material in the Chinese "Can bodies or computers have souls?" ([1993b] and [1993c]) seems to be in "On physicalism and algorithmism: Can machines think?" [1993d], as well as in "Mind, brain, machine" [1990c]. Pending further translation the authors would appreciate any evidence to the contrary.

References

Cited writings of Hao Wang:[13]

1944. The metaphysical system of the New Lixue (Chinese). *Zhe xue ping lun. Philosophical Review* 9, no. 3, 39-62.

1945. Language and metaphysics (Chinese). *Zhe xue ping lun. Philosophical Review* 10, no. 1, 35-38. Published in 1946. English translation, [2005].

1965a. Russell and his logic. *Ratio* 7, 1-34.

1966d. Russell and philosophy. *The Journal of Philosophy* 63, 670-673.

1972. Reflections on a visit to China (Chinese). *New China Bimonthly*, no. 7 (October 1), 23-26 and 31. Reprinted in *Xinwan Bao*, November 1972. Second enlarged version, *The Seventies Monthly*, no. 36 (January 1973), 54-59 and no. 37 (February 1973), 85-90; also in *Dagong Bao*, December 1972; and as a separate pamphlet by Bagu Publishing Co., February 1973. Third revised version, published as a separate pamphlet by The Seventies Publishing Company. Also in *People's Daily, Reference Information*, March 3, 4, 5, 6, 7, 1973.

1973a. Forty years of culture in Hong Kong (Chinese). *The Seventies Monthly*, no. 44 (September), 22-23.

1973b. (With Bradford Dunham.) A recipe for Chinese typewriters. IBM report RC4521, September 5.

1974c. Concerning the materialist dialectic. *Philosophy East and West* 24, 303-319.

1976a. (With Bradford Dunham.) A recipe for Chinese typewriters (Chinese). *Dousou Bimonthly*, no. 14 (March), 56-62.

1977b. The searchings of Lu Xun (Chinese). *Dousou Bimonthly*, no. 19 (January), 1-16.

1977c. Dialectics and natural science. *Overseas Chinese Life Scientists Association Newsletter* 1, no. 2 (March), 48-55.

[13]Further publication details are available in the bibliography in this volume.

24

1979b. Mechanical treatment of Chinese characters (Chinese). *Dianzi Jisuanji Dongtai*, no. 6, 1-4.

1979c. China today and its development over the last sixty years (Chinese). *Wide Angle Monthly*, no. 86, 32-49.

1981b. Remeeting Mr. Shen Congwen (Chinese). *Hai Nei Wai* 28, 25-26. Reprinted in *Dadi*, no. 2, 27-28.

1981e. Gödel and Wittgenstein (Chinese). *Philosophical Research Monthly*, no. 3, 25-37.

1981f. Mathematical logic (Chinese). *Problems of Natural Sciences*, no. 3, 70-71.

1982a. To confirm some impressions by Lu Xun (Chinese). *Dushu Monthly*, April, 70-76.

1982b. Memories related to Professor Jin Yuelin (Chinese). *Wide Angle Monthly*, no. 122, 61-63. Reprinted in *Chinese Philosophy* 11 (1984), 487–493. Also reprinted in Liu Peiyu (ed.), *The Reminiscences of Jin Yuelin and Reminiscences about Jin*, pp. 161-167. Chengdu: Sichuan Educational Press, 1995. English translation in this volume.

1983. Philosophy: Chinese and western. *Commentary: Journal of the National University of Singapore Society* 6, no. 1 (September), 1-9.

1984d. Thought and action. *South China Morning Post, The Hong Kong Standard*, June 1.

1985a. *Beyond Analytic Philosophy. Doing Justice to What We Know.* Cambridge, Mass.: MIT Press. Paperback edition, 1987.

1985b. Two commandments of analytic empiricism. *The Journal of Philosophy* 82, 449-462.

1986a. China and Western philosophy (Chinese). *Chinese Culture Quarterly* 1, no. 1 (September), 39-60.

1987b. Gödel and Wittgenstein. In Paul Weingartner and Gerhard Schurz (eds.), *Logic, Philosophy of Science and Epistemology*, pp. 83-90. Proceedings of the 11th International Wittgenstein Symposium. Kirchberg am Wechsel, Austria, 4-13 August, 1987. Vienna: Verlag Holder-Pichler-Tempsky.

1987c. On distinguishing problems of different orders (Chinese). *Chinese Culture Quarterly* 1, no. 4 (summer), 35-40.

1987d. Einstein and Gödel: Contrast and friendship (Chinese). *Journal of Tsinghua University* 2, no. 1, 32-39, 56.

1987e. The way of Jin Yuelin (Chinese). In Institute of Philosophical Research, Chinese Academy of Social Science (ed.), *Studies in Jin Yuelin's Thought*, pp. 45-50. Chengdu: Sichuan People's Publishing Co.

1987f. Review of Galvano Della Volpe, *Logic as a Positive Science* (Chinese). *Chinese Culture Quarterly* 1, no. 3 (spring), 101-104. Translated from English by Ser-min Shei.

1987g. Review of David Rubinstein, *Marx and Wittgenstein: Social Praxis and Social Explanation* (Chinese). *Chinese Culture Quarterly* 1, no. 3 (spring), 104-107. Translated from English by Ser-min Shei.

1989b. A reading of Wang You-qin on Lu Xun (Chinese). *Nuxingren (W. M. Semi-Annual)*, no. 2 (July), 118-133.

1990a. *Computation, Logic, Philosophy. A Collection of Essays.* Beijing: Science Press. Dordrecht: Kluwer Academic Publishers.

1990b. Philosophy through mathematics and logic. In Rudolf Haller and Johannes Brandl (eds.), *Wittgenstein—Towards a Reevaluation*, pp. 142-154. Proceedings of the 14th International Wittgenstein Symposium, Centenary Celebration, Kirchberg am Wechsel, Austria, 1989. Vienna: Verlag Holder-Pichler-Tempsky.

1990c. Mind, brain, machine. *Jahrbuch 1990 der Kurt-Gödel-Gesellschaft*, pp. 5-43. Proceedings of the First Kurt Gödel Colloquium, Salzburg, Austria, September 1989.

1990e. Between philosophy and literature (Chinese). *Dushu Monthly* (April), 58-66.

1991d. Gödel and Einstein as companions. In John Brockman (ed.), *Doing Science. The Reality Club*, pp. 282-294. New York: Prentice Hall Press.

1993b. Can bodies or computers have souls? I. Psychophysical parallelism and algorithmism for the physical world (Chinese). *Twenty-*

First Century Bimonthly, no. 15 (February), 102-110.

1993c. Can bodies or computers have souls? II. On algorithmism of the mind and the problem of feasibility (Chinese). *Twenty-First Century Bimonthly*, no. 16 (April), 72-78.

1993d. On physicalism and algorithmism: Can machines think? *Philosophia Mathematica* (3) 1, 97-138.

1993f. From Kunming to New York (Chinese). *Dushu Monthly* (May), 140–143. English translation in this volume.

1993g. New directions in science and in society: From traditions to innovations. In *Nineteenth World Congress of Philosophy, Moscow, August 1993, Book of Abstracts: Invited Lectures*, pp. 52-59. Moscow.

2005. Language and metaphysics. Translation by Richard Jandovitz and Montgomery Link of Wang [1945]. *Journal of Chinese Philosophy* 32, no. 1, 139-147.

Other cited writings:

Frege, Gottlob, 1879. *Begriffschrift. Eine der arithmetischen nachgebildete Formelsprache des reinen Denkens.* Halle: Nebert.

Grossi, Marie L., Montgomery Link, Katalin Makkai, and Charles Parsons, 1998. A bibliography of Hao Wang. *Philosophia Mathematica* (3) 6, 25-38. Revised and updated in this volume.

He, Zhaowu, 1995. Huainian Wang Hao. *Xi Nan Lian Da xiao you hui jian xun (Southwest Associated Alumni Newsletter)* (October), 48-49. Reprinted in He, *Li shi li xing pi pan lun ji*, pp. 779-782, Beijing: Qing hua da xue chu ban she, 2001. English translation in this volume.

Jin, Yuelin, 1937. *Luo ji.* Shanghai: Shang wu yin shu guan (Min guo 26).

Russell, Bertrand, 1940. *An Inquiry into Meaning and Truth.* New York: W. W. Norton.

Memories Related to Professor Jin Yuelin[1]

Hao Wang

Translated by Montgomery Link and Richard Jandovitz

This fall, the Institute for Philosophy at the China Social Science Academy will hold a commemorative conference celebrating "Jin Yuelin's Fifty-five Years of Teaching and Scientific Research." This is the first commemorative conference of its kind ever held on the premises of the Social Science Academy. Since I am "proud to be considered a student" of Professor Jin, I have been asked to express myself in writing. Last month I wrote a draft and sent copies to a small number of friends, asking for corrections. Some friends said what I wrote was "very interesting and should be openly published," so I have added these few lines explaining the motive force behind the draft. I hope those young friends who aspire to the arts and sciences will find a little resonance in this essay.

1 October 1982

My contact with Professor Jin Yuelin centered around our time at Southwest Associated University in Kunming. I went to Kunming in 1939 as a student and, having stayed there all the way until the fall of 1946—almost seven years, left to prepare to go abroad. During these most impressionable years, I enjoyed the kind of intimate

---

[1]This is a translation of "Cong Jin Yuelin xian sheng xiang dao de yi xie shi," i.e. Wang [1982b]. Hao Wang explains in the introductory paragraph that the present essay is in style quite informal and in preparation a rough draft. The footnotes are by the translators. Pinyin is used for the transcription of Chinese characters, except for cases in which a name is widely known in the West in its Wade-Giles form or some variant thereof. Wang writes more about his teacher Jin Yuelin (1895-1984) in [1987e]. Thanks to Charles Parsons.

and pure human relationship with many teachers and classmates that is hard to attain in life. This kind of experience not only established a relatively solid base for me later on as a man and in my academic career, but also the trust and sympathy we had for each other has always been maintained, becoming an important spiritual pillar on the rough road of life. So, when I think of Professor Jin, I think of life in Kunming, especially those teachers and friends so perfectly matched in sentiment and ideas.

With the influence of these good teachers and fine friends, paired with youthful vigor, a simple and crude material for life, the concentrated focus on scholarship and learning, and an enthusiasm for an open mind in the intellectual realm, I passed some lively days in Kunming. For many years, my strongest desire has been that the comprehensive development of Chinese society would increasingly allow more young people to enjoy this kind of rich intellectual life. Within the whole of society, the general tendency is that this kind of intellectual resource increases with use. I have always considered China's human resources, whether from the aspect of quantity or quality, to be the most abundant in the entire world. All that is needed is for the youth to get an opportunity to develop their potential in both cultural and technological aspects, and China will very quickly stride to the front lines of humanity. I think that according to the special circumstances of China, if a few more people would strive to bring about cultural development, the improvement of cultural life would proceed faster than that of material life.

Fifty-five years ago, Professor Jin returned to Tsinghua from abroad and began his academic work there. That would have been in 1927, when, like the night-blooming cereus, the Tsinghua Institute of Chinese Classics (1925-1929) was already nearing its coda. Wang Guowei committed suicide that year, and Lian Qichao was sick, and would pass away the following January, leaving only Chen Yinge from among the major instructors. (In the fall of 1945, Professor Chen passed through Kunming on his way to visit England; Professor Shen took me to pay a visit to the *great teacher*.) It was probably just before or just after that year[2] that Professor Jin began researching and promoting logic.

Among Professor Jin's students, those with whom I am most

---

[2] 1927.

familiar are Professor Shen Youding and Professor Wang Xiandiao, both of whom were also my teachers. Besides these, Mr. Ren Hua was my senior at both Tsinghua University and Harvard University. When I first got to Harvard in the fall of 1946, Mr. Ren had already finished his studies and was preparing to return home; he helped me out quite a bit. I've recently seen some materials saying that Professor Jin graduated from the preparatory department for study in America in 1914; Professor Shen graduated in 1929, in which year Tsinghua changed to a university; Professor Wang graduated in 1933; Mr. Ren graduated in 1935. In 1933, Tsinghua established a research institute, and in ten years time had only three graduates: Mr. Ren in 1937, Zhang Daowu in 1940, and me in 1945. It just occurred to me that the Chinese students who have studied logic at Harvard include only Mr. Zhao Yuanren, Mr. Yu Dawei, and me. Both Zhao and Yu quickly changed their occupations, and I, too, always have felt that logic is a bit too narrow.

Professor Jin had another student called Yin Fusheng, who later changed his name to Yin Haiguang. Master Yin had no understanding of logic whatsoever, but he liked to promulgate it. He was good at public speaking and writing articles promoting independent thought; in the 1950s and 1960s he had quite an influence on young folks in Taiwan and Hong Kong. He mentioned Professor Jin in many venues; moreover, he considered it to be his greatest honor and glory to have been one of Professor Jin's students. After Master Yin died, one of his students came out with a book, *Spring Silkworms Spitting Silk,* which included many things concerning Mr. Jin;[3] one of them tells of the Logic Research Group in Beijing (then called Peiping) before the War of Resistance. During one of the meetings, someone brought up the importance of Gödel's[4] work:

> Jin Yuelin said he was going to buy himself a book to read up on. His student Shen Youding said to Professor Jin, "Truth be told, you wouldn't understand it." Professor Jin heard this, interjected two sounds, "eh, eh," then said, "Then forget it." Yin Haiguang was to one side observing this conversation between teacher and student and caught

---

[3]Yin [1979].

[4]The Chinese name Wang uses is uncharacteristic and unfamiliar to us. It could very well be a name for Kurt Gödel, but in extensive discussions of the logician elsewhere Wang never uses it, to our recollection.

30

his breath in surprise. A student's criticism, a comment
which the teacher immediately accepted, was unprecedented
in China.[5]

I think that this story is most likely not made up; furthermore, I
feel everyone should have Professor Jin's measured refinement; if
in a society this kind of reasonable response is seen as strange and
wonderful, then that is really a shame.

In 1938, while I was in high school, I saw *Logic,* the book Professor Jin had written, which seemed like a lecture handout; this
was not the college reader put out by the Commercial Press toward
the end of 1963.[6] In elementary school, I had read the textbook my
father wrote on ethics and felt it was completely without content.
Later, my father wanted me to read "Anti-Schelling" and *On Feuerbach,* but I simply could not understand them.[7] Yet, when I read
Professor Jin's book, especially the chapter on mathematical logic,
I felt it was very easy and thought perhaps I should start studying
this kind of simple stuff; later, when I became more mature, I could
go back to the study of dialectics.

The next year I entered Southwest Associated; because I was
late arriving on campus, I wasn't able to get into the A section
of Universal Logic, taught by Professor Yin. I was put in the B
section, taught by Zhang Yinlin. In addition, I audited the Symbolic Logic course Professor Wang Xiandiao taught for fourth-year
and graduate students. Each week Professor Jin held an evening
guided review session, which I went to once and asked some questions. I remember Professor Jin said that since I was already adept
at everything, I could skip attendance. In Professor Zhang's class,
I often pointed out errors and gave proofs, so he invited me to his
home to chat and lent me two pretty old extracurricular readings. I
remember Professor Zhang said he could not figure out the concept
of *variable*: But did he not understand what he was looking for all
along?

---

[5]Yin, *op. cit.*, p. 74.

[6]According to Wang [1987e], Jin's *Logic* was originally published in [1937]
as a Qinghai University textbook. Jin [1961] is a republication for which the
author added a criticism of the original.

[7]Wang mentions works by Engels dating from late 1841 to early 1842. We
do not have further information on the Chinese translation Wang's father had
in mind, nor on Wang's father's work on ethics. *On Feuerbach (Feierbaha lun)*
is Engels [1932].

In my second year of college, I had already finished all of my courses in mathematical logic and agreed with Professor Shen and Professor Wang that we would read the two volumes of Hilbert and Bernay's *Grundlagen der Mathematik.*[8] We worked it out that each of us would in turn be responsible for reporting on a chapter. What a pity that, after I reported on the first chapter, the two teachers didn't follow up; because of this I also became lazy and didn't continue reading. (It wasn't until seven or eight years later, when I was going to teach a class on that subject, that I finished reading the whole book.) At that time, Professor Chen Shengshen proposed that if I wanted to do a class on mathematics, all I had to do was find a comprehensive book and read it through thoroughly; then I could do research. At that time, *Grundlagen der Mathematik* just so happened to be the book that from the standpoint of mathematical logic met this standard. That I hadn't finished reading it earlier was really quite a shame.

In my third year of college, I began to take some philosophy courses. At that time, there was a kind of shallow way of thinking that considered mathematics to be only a lesser way, not equal to philosophy in importance. The chair of the Mathematics Department, Yang Wu, was extremely kind to me. He urged me to take advantage of my youth to study mathematics; philosophy could be left for later study. In my youthful arrogance, I didn't accept Professor Yang's counsel, but rather put most of my energy into philosophy. Besides reading books, I also began to focus on Hume's "problem of induction," spending more than a year of labor in writing a long paper. Later, when I submitted it to Professor Jin, he read it and asked me if it had been plagiarized. After I informed him it hadn't, he said it ought to be published. Too bad both he and I were too lazy and never exerted any effort toward that end. Now this "virgin work" cannot be found.

One time Professor Jin devoted a course solely to Bradley's *Appearance and Reality.*[9] I took it: there were only two students; the other one it seems to me now was Xu Xiaotong. We each had a copy of the book; Professor Jin would read and read, and lecture and lecture. Another time Professor Jin lent me two chapters of the

---

[8]Hilbert and Bernays [1934] and [1939].
[9]Bradley [1893].

book he was working on, *On Knowledge*:[10] later on he had forgotten
and, unable to find them anywhere, thought they were lost forever.
When I returned them to him, it was like a great burden had been
lifted. When I was writing my thesis, Professor Jin wanted me to
read Price's *Perception*.[11] Later he also recommended it to Wang
Zigao, who came to get it from me, since I knew him.

After I had graduated from the Mathematics Department undergraduate program, I tested into Tsinghua's Philosophy Department
as a graduate student. My good friend He Zhaowu graduated from
the History Department and, by passing the entrance exam, accompanied me to graduate study in the Institute of Philosophy; after
half a year, he found it wasn't to his liking and made a change by
entering the History Department at the Institute. *Elder Brother* He
was interested in intellectual history; if he hadn't entered Tsinghua
at that time but had gone to Peking University instead, maybe he
wouldn't have needed to change departments.[12] The theory at that
time was that Peking University stressed the history of philosophy,
but Tsinghua stressed the problems of philosophy. My good friend
Wang Jinghe of the Physics Department took Professor Feng Wenqian's History of Western Philosophy: Professor Feng saw him as
a philosophical genius. Elder Brother Wang did not later switch to
philosophy, but pursued mathematics as his main work.

For the two years I was at the Institute, I worked on the one
hand as a full-time middle school teacher and on the other hand as
a full-time graduate student. Besides classes, I also had a second
foreign language, along with a paper equivalent in requirements to
a western doctoral dissertation: my life was pretty intense. (My
especially outstanding colleague in the Physics Department, Zhang
Chongyu, taught with me at the Tian Xiang Middle School; it
wasn't long before he had come down with consumption; he never
recovered and died several years later. My colleague Lu Zhongrong,
with the highest marks in the Chemistry Department, met with a
similar fate. These are the most obvious examples of the waste
of human talent; each time I think of it, I get "choked up" over
it.) Because of this kind of training, when I got to Harvard it was

---

[10]Jin [1983]: the original draft dates from 1948.

[11]Price [1932].

[12]Wang writes more about He in [1993f], newly translated nearby in the
present volume.

like repeating a grade, all I had to cope with being some exams and writing another thesis. That's why it took me only fifteen months to complete the requirements for the Ph.D., not to mention publishing several articles.

At Tsinghua, Professor Jin was my thesis advisor, but I didn't see much of him, just giving him each chapter after I had finished writing it. My thesis had four chapters: one chapter was previously mentioned, "The Problem of Induction"; one chapter was "Truth"; one was "The Nature of Verifiable Experience"; the other, it seems, was "Sense Data." In those days I was like a new-born calf: when I wrote I rarely restrained myself, not like later, when I was more careful, and my writing had a completely different flavor. Too bad all these years later I've never found a copy (at the time I asked someone to write out a copy for me, so originally there were two copies): the Tsinghua Library has also been unable to turn up a copy. I remember Professor Shen once asked me if the four chapters were four articles or one book. I said they were just four articles. Later on when writing books, I was still unable to change this bad habit:[13] a book always seemed like a collection of essays. This most likely cannot be separated from a certain style of contemporary philosophical research.—All the professors participated in my thesis defense: besides Professor Jin, it seems that Feng Wenqian, Yang Yongshenme, Fung Yu-lan, Zheng Xin, He Lin, Shen Youding, and Wang Xiandiao were the professors there.[14]

When I finished up in the spring of 1945, an introduction from a teacher led me to Yunnan University to be an instructor. When I got to Chancellor Xiong Qinglai's waiting room, I laughed nervously; it seems that I also met him once or twice, but nothing came of it. Luckily, Yang Zhenning went abroad and I filled his position as a teacher of mathematics at Associated University's attached middle school. The vogue then was to go abroad, and what I was studying was of the West, so Professor Jin recommended that I apply for a scholarship to the University of Chicago, but I wasn't successful. In the spring of 1946, there was suddenly a government scholarship that I unexpectedly won, and in September I left

---

[13]Wang calls it a sickness: *gai bu liao zhei ge mao bing.*

[14]Wang's notable committee includes among other luminaries Fung Yu-lan, for whom see the translation in this volume of Wang, "From Kunming to New York," n. 5.

the country. Before I left, Professor Jin took me to visit Professor Lin Zhengyin: Professor Lin's highly refined style and manner of speaking made a deep impression.

After 1949, Professor Jin wrote several articles of self-criticism, two of which were widely disseminated abroad. One of these mentions three of his students who hadn't been able to keep up with the times: one was Professor Shen; one was Yin Fusheng; I was the other. The point was that the circumstances of these students reflected his past mistaken attitude toward learning. It was probably around this time that my father wrote to Professor Jin and Professor Fung Yu-lan, upbraiding them for leading me off the "straight and narrow," due to which I was slow in returning home. I guess what my father meant was: originally I could have studied science, but, because I was confused by them, I went into academic philosophy. Recalling the situation at college, there is some truth in this because, at that time, I did indeed feel that the work of Professor Jin and Professor Fung held more attraction than the work of Lian University's best mathematics and physics professors. Later I discovered that my ability in philosophy was not up to my ability in science, so that the work I did in science without much effort was more accepted in the academic world than all my strenuous efforts in philosophy.

For several years after 1949, because I felt what I was studying was of no use to my country, I thought I'd study something a little closer to the training I had received: calculating machines. In 1955 and 1956, there was a movement for the return of overseas students; I remember corresponding with Professor Jin about this. Because it was somewhat difficult to return to China from America, I looked for a situation in England. In 1956, I accepted a teaching and research position in the philosophy of mathematics at Oxford University. I began work in the fall of 1956. Not long thereafter, I received a letter from Chancellor Ma Yinchu of Peking University offering me a professorship; I could only politely thank him, one of the reasons being that it was not convenient to resign immediately. The anti-rightist campaign and the Great Leap Forward began two years later. Professor Ma, not long thereafter, resigned his position as chancellor.

In the spring of 1958, Professor Jin participated in a cultural delegation visiting England that stayed at Oxford a few days. It had already been twelve years since last I had seen him in Kunming. I arranged for Professor Jin to give a short report to a meeting of Oxford philosophy professors: he said that because Marxism had saved China, he had abandoned the academic philosophy he had done before and become a Marxist. Most of the professors who heard the lecture felt his proof a bit too simplistic. But because Professor Jin's British English was especially elegant and polished, the majority of Oxford professors treated him with the utmost respect. I remember over-confidently saying to Professor Jin that I hoped to expound the philosophy of Marxism a little more lucidly, and Professor Jin replied, "It's probably like this," that it was impossible to explain Marxism as lucidly as "the standard I imagined in my mind's eye." While Professor Jin was at Oxford, I drove a beat-up jalopy to pick him up and drop him off; he expressed astonishment that I was able to drive a car.

In the summer of 1972, I went back for the first time since leaving the country and went through quite a runaround[15] before seeing Professor Jin. Each time I returned home to China after that, I paid a visit to Professor Jin. In 1978, I gave him the book that I had expanded from my lectures at the Chinese Academy of Science, *Popular Lectures on Mathematical Logic,* and asked Professor Jin to write the title for the Chinese edition;[16] it was published in 1981, imprinted with Professor Jin's characters, but there was no mention that he had written them: I hope, when it is reprinted, that this mistake can be corrected. In 1980, the first issue of *Social Sciences in China*[17] included an essay by Professor Jin introducing Chinese philosophy; when I read it, it was as if I were already acquainted with it: it was originally written by Professor Jin in Kunming in 1943, which I had read then in a mimeographed form. It occurred to me that Professor Jin's thirty years of effort studying Marxism had to have changed his views, and so probably, because he was getting on in years, he was unable to rewrite it.

In June of this year I saw Professor Jin: he was in the process of writing his memoirs; on his desk was a thick pile of manuscripts; I was quite satisfied to see this. Pity was that his hearing had deteriorated; I hope that he is able to get a prescription for a suitable

---

[15] *Fei le xie zhouzhe*, with the *zhou* typographically in error.

[16] Wang [1981a].

[17] *Zhongguo she hui ke xue.*

hearing aid so that next time we talk it will be a little easier. Unfortunately I cannot attend in person this 55th Anniversary Commemorative Meeting, so rich in meaning, but I hope it will be permitted that my reminiscences can retain their original shape and be read aloud.

1 September 1982

References

Cited writings of Hao Wang:[18]

1981a. *Popular Lectures on Mathematical Logic.* Beijing: Science Press. Reprinted with a Postscript, New York: Dover, 1993.

1982b. Memories related to Professor Jin Yuelin (Chinese). *Wide Angle Monthly,* no. 122, 61-63.

1987e. The way of Jin Yuelin (Chinese). In Institute of Philosophical Research, Chinese Academy of Social Science (ed.), *Studies in Jin Yuelin's Thought,* pp. 45-50. Chengdu: Sichuan People's Publishing Co.

1993f. From Kunming to New York (Chinese). *Dushu Monthly* (May), 140-143. English translation in this volume.

Other cited writings:

Bradley, F.H., 1893. *Appearance and Reality: A Metaphysical Essay.* London: S. Sonnenschein.

Engels, Friedrich, 1932. *Feierbaha lun.* Translation into Chinese by Jiasheng Peng of *Ludwig Feuerbach und der Ausgang der klassischen deutschen Philosofie* (reprinted, Berlin: Dietz Verlag, 1952). Shanghai: Nan qiang shu ju.

Hilbert, David, and Paul Bernays, 1934. *Grundlagen der Mathematik,* volume 1. Berlin: Springer.

Hilbert, David, and Paul Bernays, 1939. *Grundlagen der Mathematik,* volume 2. Berlin: Springer.

Jin, Yuelin, 1937. *Luo ji.* Shanghai: Shang wu yin shu guan (Min guo 26).

---

[18]For a complete list in greater detail of Wang's writings, see the bibliography in this volume.

Jin, Yuelin, 1961. Revised edition of Jin [1937] with criticism added. Beijing: Commercial Press. Reprinted, 2002.

Jin, Yuelin, 1983. *Zhi shi lun*. Beijing: Commercial Press.

Price, H. H., 1932. *Perception*. London: Methuen.

Yin, Haiguang, 1979. *Chun can tu si*. Taipei: Yuan jing chu ban she (Min guo 68).

From Kunming to New York[1]

Hao Wang

Translated by Richard Jandovitz and Montgomery Link

During the War of Resistance, He Zhaowu and I were classmates for many years.[2] Among close friends at Southwest Associated University, names of more than one character were often shortened; however, rather than following the usual convention of leaving off the surname, we called the author "He Zhao," omitting the character *wu*.[3]

Although I have an amateur interest in Western historiographical thought and Sino-Western *cultural intercourse,* I have never been a serious student of them, so He Zhao has allowed me to write here as I please. I would like to recall a few old matters that pertain to the both of us so as to reflect the experiences a generation has had studying the liberal arts.—Besides the many years in Kunming during which I often saw him, in 1980 and 1984 He Zhao came to New York, [where] we spoke frequently.

This more or less describes the kind of person I have in mind:

---

[1]This is a translation of Wang [1993f], an essay first published as a preface to a collection of critical essays by He Zhaowu: please see the editorial comment following this essay and the adjoining footnote. Please note that all footnotes have been added by the translators. Please note also the policy of the translators here for the transcription of Chinese characters: pinyin shall be used except in those cases where a name is widely known in the West in its Wade-Giles form or some variant thereof. This leads to the following sort of divergence: 'Wei Duanji' but 'Fung Yu-lan'. Sometimes a footnote offers the variant. Thanks to Charles Parsons.

[2]The War of Resistance against the Japanese, 1937–1945; He (1921–) is also known as Ho Chao-wu.

[3]The Chinese character denoted by *wu* has a warlike aspect. Southwest Associated is Xinan Lian in Kunming, a wartime combination of Peking University, Tsinghua University, and Nankai University.

Born into the post-May Fourth Movement[4] China, primarily interested in thought and theory, lacking the motive and the ability for practical action, and moreover reluctant to give up the quest for an ideal, unifying knowledge. The circumstances of each person are, in the grand scheme of things, largely identical although having their little differences. For example, when I was twenty-five I left the country to study abroad, while He Zhao stayed in China: since then our circumstances have differed greatly.

The general mood of the thirties held that engineering was supreme: For the most part, the middle school students with the highest marks planned to enter engineering academies. In 1939, He Zhao passed the entrance exam and matriculated into the School of Engineering at Southwest Associated; but, because he had grown up in the ancient city of Beijing and, moreover, had encountered a period of warfare fraught with change, the next year he set his heart on transferring to the Department of History. That same year I entered the Department of Mathematics, for I was hoping to study philosophy and thought of mathematics as a foundation. After completing the basic course work, we passed the entrance exam and matriculated at the same time into the Tsinghua University Institute of Philosophy; there, He Zhao soon discovered the subject matter did not meet his expectations, and so the next year he transferred to the Institute of History.

From our time as graduate students I still remember several things all these years later. Once, He Zhao got hold of a record player and some albums. In his room I came into contact, for the first time, with Western classical music, which I felt quite pleasant to hear. I still remember listening to Mozart's "Eine Kleine Nachtmusik." Later, when I got to America and told a foreign classmate that I liked this piece, he seemed to feel my taste was not sufficiently refined.

We all used to like reading a mixed lot of books, occasionally even quoting some lines. One time He Zhao said, "Russell has a dry humor, while Freud has a wet wit." Another time, he related three kinds of love: animal attraction, heat-seeking, and mutual creation. His gist was that it is best to have all three.

There were a couple of things I had told He Zhao then that,

---

[4]A reaction against the West and Japan that occurred after World War I.

when he brought them up again many years later, revealed to me that my understanding of what two of my professors had said had been woefully inadequate.—In 1945, during the defense of my master's thesis, Professor Shen Youding asked me why I wanted to study philosophy. I said it was because I was interested in the human dilemma. Professor Shen said, "In the West it is literature that focuses on the human dilemma, not philosophy." Another time Professor Fung Yu-lan[5] said to me, "Everyone familiar with both Chinese and Western music prefers Western music; everyone understanding both Chinese and Western philosophy prefers Chinese philosophy."

After 1949, He Zhao did a lot of translation work, including works of Russell, Rousseau, Pascal, Kant, Greenwood, Needham, Popper, Meinecke, and others.[6] Early in the 1950s, my father hectored me to return home, saying after I returned to China I could go to work at some translation bureau. Then, moreover, he wrote letters to Professor Jin Yuelin and Professor Fung Yu-lan, upbraiding them for leading me astray from the universal applicability of mathematics to idealistic philosophy.[7] At the time, I felt that, given what I was studying, returning home would be useless, especially since, besides mathematical logic and Anglo-American philosophy, I was beginning to do a little research into work of a similar nature, computation theory.

---

[5]1895–1990, also known as Feng Yu-lan and Feng Youlan. Fung did his undergraduate work at Peking University and earned his Ph.D. at Columbia University. Fung was a professor of philosophy at Tsinghua University beginning in 1928 and was dean of Southwest Associated University while Wang and He attended there. Later Fung became a professor at Peking University. He is perhaps the most influential Chinese philosopher of the twentieth century and best known for his [1952], first published in Chinese in 1931, and [1953]. Among his works in original philosophy must be mentioned *The New Rational Philosophy* [1939], which indicates his ability as a systematic philosopher; later, in the fifties, he seems to have retracted the new rational philosophy. For further information, see Chan [1963], 754–755, and for Fung's own account of his work, Fung [1948], ch. 28.

[6]The People's Republic of China was founded in 1949.

[7]About Jin (1895–1984), also known as Chin Yüeh-lin, Wing-tsit Chan has written that this "expert in logical analysis," "much influenced by T. H. Green, has developed his own system of logic and metaphysics based on it" [1963], 744. For more on Jin see Wang [1982b], translated in this volume, and Wang [1987e].

42

In 1956, He Zhao began working at the Institute of History,
China Academy of Science, where he participated in the compilation and writing of *A General History of Chinese Thought* and
several other such books.[8] Before the Cultural Revolution,[9] He
Zhao published a few essays, his translations of two books by Russell introducing philosophy and the history of western philosophy,
as well as Rousseau's *Social Contract* and *Discourse on the Sciences
and the Arts*.[10]

After the Cultural Revolution, He Zhao gradually published the
great bulk of his essays and translations.[11] As a result of his accu-
mulated experience and scholarship, plus his diligent work habits,
the quality and quantity [of his work] increased greatly year by
year. Besides essays and translations, he also revised a dozen or
more books, wrote *An Intellectual History of China* in English, and
translated *The Wisdom of China*.[12]

He Zhao's thought has up to now been disseminated for the
most part in the form of essays. Those essays treating of Chinese
thought have already been variously anthologized. Now this book
collects his scattered works commenting on Western historiographical thought and Sino-Western cultural intercourse.[13] He is currently
planning to write several systematic books.—I consider the course
of the development of his work to represent a quite healthy evolutionary path in the study of the liberal arts.

The course of my own development as well has been from scattered essays to collected articles to relatively systematic books, but
I have always felt that writing books was more difficult than writing articles, the degree of difficulty out of proportion to the length.
Yet, the greatest difficulty I have encountered still is the inability
to synthesize my specialized work and my philosophical ideals.

Beginning in 1966 or so, I decided to focus my attention mainly
on philosophy. I gradually came to realize, just as Professor Shen
had said that year, that it would be quite difficult to use my previous work as a foundation and add to it my concerns about the

---

[8]He [1960].

[9]1966–1976.

[10]Russell [1960], [1963]; Rousseau [1962], [1959].

[11]See the list of He's writings in He [1998], 446–450. No publications are
listed for He during the years 1964–1977, i.e. the Cultural Revolution.

[12]He et al. [1991].

[13]He [1994]. See the editorial comment following this essay.

problems of human society, taking my ideas in two different fields and integrating them into one all-inclusive, comprehensive point of view.—When I first went to college, I considered Russell to have successfully integrated his two disparate interests; only later did I know his work in these two areas in practice was done each in its own way.

In the summer of 1972, twenty-six years after I had left the country, I returned to China for the first time and became fascinated by the Marxist world view. Although there was much that puzzled me, I felt that I saw a comprehensive philosophy unifying thought and action in a way that I had not before considered possible.— During my middle school years, my father wanted me to read some books about dialectics and materialism, which at the time I felt I did not understand. Later, in the third year of high school, having read Professor Jin Yuelin's textbook, *Logic,* I felt mathematical logic was easy to understand.[14] I thought if I first studied what was easy, later I might perhaps be able to understand what was difficult.

My infatuation with Marxism, which began in the fall of 1972, lasted about seven years; yet, I had felt all along that my knowledge of history and contemporary fact was inadequate; as a result, I did not have the ability to judge which views passed the test of fact. From 1977 on, a Chinese glasnost reevaluated a certain number of the views of the Cultural Revolution and earlier. In particular was the revelation, from private conversations and literary works, of many concrete facts about which most persons before had had no idea. After coming across such information, I began to doubt my own most recent views. I felt that, due to my deeply rooted and stubborn patriotic thinking, I had imagined certain facts that were grounded in wishful thinking; upon this foundation I had erected a logical construction that really was a castle in the sky.

For three years or so, I slowly transformed my way of thinking, bringing to fruition some doubts and suspicions about those Marxist positions closest to my heart, especially those concerning concrete applications, about which a number of questions had appeared: my

---

[14]Jin [1937]. The 1961 edition starts with a self-criticism by the author, which seems to acknowledge that the original was shot through with the logical thinking of the capitalist class and expresses regret that this may have had a baleful influence.

44

heart was unusually bitter and depressed. It was probably not until '81 that I began to acknowledge that my training and talent were not suited to the study of such a great philosophy as Marxism, that it would be best to put to good use the stuff with which I was most familiar, to do some work brick by brick. I still hoped to draw out of Marxism some essence I could digest. Simultaneously, I began to study and think some about traditional Chinese philosophy. Whenever I feel that my own philosophical work is unable to mesh with my concerns about the human dilemma, I have a deeper appreciation than before for what Professor Fung said about the drawing power of Chinese philosophy.

In 1980, He Zhao came to New York; we had many opportunities to see each other.[15] I remember the pain and bitterness I suffered at the time because of the wrenching transformation of my thinking and the loss of my convictions; more often than not I grumbled a lot. He Zhao before returning home transcribed and presented me with the poetic pentalogy, *Bodhisattva Barbarians,* by Wei Duanji. Therein, Wei wrote of the five stages in the evolution of his own state of mind after leaving his native land.[16] At the time, I thought I understood the gist, but many of the subtleties eluded my grasp. Later, after reading Mr. Yu Pingbo's interpretation of it in *Du Ci Ou De,* I reaped a more bounteous harvest.[17] I even wrote an essay comparing my homesickness in a foreign land with each of the stages from the five poems. Later, I felt the essay to be sick with self-pity and, thinking it nauseating, I did not publish it. Consequently, my admiration for poets has been strengthened.

With the dominant status of technology and commerce in the twentieth century, not only in China but also elsewhere, it seems as if all persons with more or less the same interests as He Zhaowu and I must walk this or that kind of crooked path. Misery loves company, so it is only natural that I should have a special fellow-feeling for this kind of person; it is for this kind of person that I consider Lu Xun's[18] advocacy of tenacity to be especially important.

---

[15]He went to New York City as a scholar with the Sino-American Cultural Intercourse Committee.

[16]Wei Duanji is also known as Wei Zhuang, 836–910, an innovator in the development of *ci* (song poetry) in the late Tang dynasty.

[17]Yu (1899/1900–1990) [1959].

[18]1881–1936.

In 1984, He Zhao stayed in New York for the better part of the year, and we had many joyous get-togethers.[19] I was right in the middle of writing a book on philosophy then;[20] I considered some of the problems to be of a relatively general nature and very much wanted to discuss my way of thinking with him, but he seemed to continue feeling that it was all too specialized. Because of this I realized that I had always lacked the ability, outside my specialization, to express my thoughts effectively, even though I had exerted much effort in this direction for many years. Besides, maybe there is in China a relative lack of the habit of frank and earnest discussion and mutual criticism: Don't be afraid to ask questions; don't be afraid that you perhaps are wanting in some relevant knowledge.— Recently I discovered that telephoning Beijing from New York has become quite easy. So, I often have the opportunity to chat with He Zhao on the phone. This way, we save a lot of trouble writing out characters. One of the things we agree upon in our conversations is that both of us hope genuine academic criticism can develop in a healthy manner, never again to be punitive denouncement in the style of mass criticism, nor the flattery of current fashion, but a discourse that is a genuine boon to the exchange of academic ideas.

23 December 1992

He Zhaowu is a professor in the Institute of Ideology and Culture at Tsinghua University in Beijing. For many years he has studied western culture and has been a frequent contributor to these pages. This is the preface the American professor Hao Wang wrote for the latest work of He Zhaowu, Critical Essays on History and Reason.[21]

---

[19]He was a visiting professor at Columbia University, 1983–1984.

[20]Wang is most likely referring to [1985a].

[21]This last paragraph was added by the editor of *Dushu Monthly*. He's work mentioned by the editor is [1994].

46

References

Cited writings of Hao Wang:[22]

1982b. Memories related to Professor Jin Yuelin (Chinese). *Wide Angle Monthly,* no. 122, 61-63. English translation in this volume.

1985a. *Beyond Analytic Philosophy. Doing Justice to What We Know.* Cambridge, Mass.: MIT Press.

1987e. The way of Jin Yuelin (Chinese). In Institute of Philosophical Research, Chinese Academy of Social Science (ed.), *Studies in Jin Yuelin's Thought,* pp. 45-50. Chengdu: Sichuan People's Publishing Co.

1993f. From Kunming to New York (Chinese). *Dushu Monthly* (May), 140-143.

Other cited writings:

Chan, Wing-Tsit, 1963. *A Source Book in Chinese Philosophy.* Princeton: Princeton University Press.

Fung, Yu-lan, 1939. *Xin li xue.* Changsha: Shang wu yin shu guan.

Fung, Yu-lan, 1948. *A Short History of Chinese Philosophy.* Derk Bodde, ed. New York: Macmillan, 1948.

Fung, Yu-lan, 1952. *A History of Chinese Philosophy,* volume I. *The Period of the Philosophers (from the Beginnings to circa 100 B.C.).* Translated by Derk Bodde. Second edition. Princeton: Princeton University Press. Original publication. Peiping: Henri Vetch, 1937.

Fung, Yu-lan, 1953. *A History of Chinese Philosophy,* volume II. *The Period of Classical Learning (from the Second Century B.C. to the Twentieth Century A.D.).* Translated by Derk Bodde. Princeton: Princeton University Press.

He, Zhaowu, 1960. *Zhongguo si xiang tong shi.* Beijing: Ren min chu ban she.

---

[22]For a full list in greater detail of Wang's writings, see the bibliography in this volume.

He, Zhaowu, 1994. *Li shi li xing pi pan san lun*. Changsha: Hunan jiao yu chu ban she.

He, Zhaowu, 1998. *He Zhaowu xue shu wen hua sui bi*. Beijing: Zhongguo qing nian chu ban she.

He, Zhaowu, Bu Jinzhi, Tang Yuyuan, and Sun Kaitai, 1991. *An Intellectual History of China*. Translation with revisions by He Zhaowu of *Zhongguo si xiang fa zhan shi* (Beijing: Zhongguo qing nian chu ban she, 1980). Beijing: Foreign Languages Press.

Jin, Yuelin, 1937. *Luoji*. Shanghai: Shang wu yin shu guan (Min guo 26).

Rousseau, Jean-Jacques, 1959. *Lun ke xue yu i shu*. Translation by He Zhaowu of *Discours sur les sciences et les arts*. Beijing: Shang wu yin shu guan. Revised, 1963.

Rousseau, Jean-Jacques, 1962. *She hui qi yue lun*. Translation by He Zhaowu of *Du contrat social*. Beijing: Shang wu yin shu guan.

Russell, Bertrand, 1912. *The Problems of Philosophy*. London: Williams and Norgate. Reprinted, Mineola: Dover, 1999.

Russell, Bertrand, 1960. *Zhe xue wen ti*. Translation by He Zhaowu of Russell [1912]. Beijing: Shang wu yin shu guan.

Russell, Bertrand, 1963. *Xi fang zhe xue shi*. Translation by He Zhaowu of *A History of Western Philosophy*. Beijing: Shang wu yin shu guan.

Yu, Pingbo, 1959. *Du Ci Ou De*. Hong Kong: Xianggang wan li shu dian.

Remembering Wang Hao[1]

He Zhaowu

Translated by Richard Jandovitz and Montgomery Link

Wang Hao has unexpectedly left this world and gone; where did he go?[2] To that "from each according to ability, to each according to need" real world of Marx? Or to that paradise within this mortal world, the fleeting moment in the flow to which Goethe wants to call out, "Ah, still delay—thou art so fair!"?[3] Or to Plato's world of eternal ideals? For many years he has longed for all of these worlds.

Surveying Brother Hao's life—except for that period of material hardship when he was a student during the War of Resistance (which at the same time was one of an unusually fertile spiritual life)—one could say it was smooth sailing, with success and fame. Yet his whole life from beginning to end was "chock full of contradiction," be it in his thought, in his studies, or in his personal life. These contradictions were not only his own, they belonged also to a part of the difficult course of an entire generation and people.

Born to a family of intellectuals, in his youth and adolescence he always did exceedingly well in the middle schools and universities with the best reputations in China at the time. Later on he completed his graduate work with the greatest alacrity and, on the recommendation of his alma mater, Tsinghua University, won a U.S. State Department scholarship to attend Harvard University.

[1]The Chinese original is He [1995]. Translated here by permission of the author and the editor of the *Southwest Associated Alumni Newsletter*. The translation is published here to complement that of Wang [1993f]. The footnotes are due to the translators. Thanks to He Zhaowu for his gracious help. Thanks also to Charles Parsons and Zhu Wenman.

[2]Recall that He wrote this piece some fifteen years ago.

[3]"Verweile doch! du bist so schön!" *Faust*, part 1, line 1700, Bayard Taylor translation.

Studying during that time under the famous professor Quine, he again needed only one year and eight months' time to earn his Harvard Ph.D. After this, achievements, honors, and status followed closely on his heels. He was publicly acknowledged as the inheritor of the mantle of Gödel, the greatest mathematical logician and philosopher since Leibniz.[4] Concurrent with the decades of teaching and research work he held a position at IBM. Although his income was considerable, throughout his life he was not good at managing money matters, so it is unlikely he had much in the way of savings.

He made groundbreaking contributions in the field of mathematical logic, making him a world-class authority already at an early age. But in those one-sided political days of the 1950s mathematical logic was denounced by the Soviet Union as a conceptual game of capitalist class idealism, a calamity extending even to the fish in the moat.[5] Very much wanting to return home at that time and do a service to his country, he changed his field of research to computing machines, thinking to do something that would be of some use after he returned to China. Peking University Chancellor Ma Yinchu wrote him offering him a teaching position, but he was still emotionally attached to philosophy.[6] His philosophical approach was "incompatible with the needs of the time" of the contemporary world of Anglo-American philosophy, but he was unwilling to abandon his own approach and take up the analytic philosophy then in vogue, even though had he done so it was certain he would have met with outstanding success. One opportunity after another slipped away right up until the Cultural Revolution, at which time it became impossible to return.

Politically he was a leftist—at least outside China. In 1972, after China and the U.S. broke their deadlock, he participated in the first delegation of overseas scholars to return to China. After this he dove headlong into Marxism, persevering steadfastly, making the best of the years, until after the Gang of Four was smashed and the darkness of the Cultural Revolution was gradually exposed. This

---

[4]He reports that in China Wang has been seen as the inheritor of Gödel's mantle, a public acknowledgement that would not extend globally.

[5]Research in mathematical logic was pursued, as part of mathematics, in the USSR.

[6]Ma (1882-1982) was chancellor of Peking University (1951-1960). His "New Population Theory" predicted catastrophic population growth.

caused overseas leftists to fall into awkward circumstances hard for us in China to imagine. (In China, for instance, one probably need not worry about being interrogated as to why one supported something yesterday and shouted "Down with it!" today.) At this time he suffered a severe attack of disillusionment. His morale at an all-time low, instead of turning away and detaching from reality, he engaged in an inquiry into pure philosophy, a relatively comprehensive inquiry into the fundamental problems of philosophy. Up until his departure from this world he completed three works: *Beyond Analytic Philosophy* [1985a], *Reflections on Kurt Gödel* [1987a], and *A Logical Journey: From Gödel to Philosophy* [1996a]. In addition, his early book, *From Mathematics to Philosophy* [1974a], has been translated into many languages, but although it has been more than ten years the Chinese manuscript has yet to be seen by readers in China.[7] In speaking of this last year, he expressed great regret.

Markedly different than the rough and bumpy road traveled by many intellectuals of the same period in China, who suffered criticism and denunciation, were unable to work at their professions, often losing their skills entirely, accomplishing nothing, entire lives discarded, Wang Hao's life overseas seen from the surface seems smooth and steady over the course of the decades as professor at England's Oxford University and America's Harvard and Rockefeller Universities. But contradiction and anxiety still accompanied him throughout his life. Those outstanding achievements that earned him world fame just so happened to come from aspects of his work that came easy to him. He was always thinking to take logic as a point of departure from which to build a systematic philosophy. In his later years he had already gradually abandoned this ambition, but still always bore in mind the intent to explain the three fundamental problems of human life. That he experienced many waves in his emotional life is something his old friends know very well.

Philosophy in the final analysis is the study of human knowledge. This not only requires the use of instrumental rationality, but also requires the total commitment of one's thought and emotions. In the operation of pure reason he proved himself to be one

---

[7]Since the only complete translation that has appeared is in Italian, the statement that this work has been "translated into many languages" is not accurate.

52

of the great masters of the modern world, but when he needed to use his spirit to wrestle with the true meaning of life he often displayed naïveté or immaturity, sometimes seeming like an innocent yet headstrong child. He wanted a comprehensive philosophical system to embrace the world and human life, but in the end (as he himself in his later years perceived) it was perhaps just as Hamlet says to Horatio:

> There are more things in heaven and earth, Horatio,
> Than are dreamt of in your philosophy.[8]

In his youth he liked to examine minutely: what is happiness? He cited Gide: people are born for happiness. If not happiness, then what else would it be? What people seek is happiness, not glory, knowledge, power, status, or even the sublime or holy or any other thing; although he acknowledged that these have a connection with happiness, people often must get past glory before being able to attain peace of mind. But even in his later years he seemed not to have found the kind of happiness he had sought after in his youth. Of course, academic research, friendship, love, nationality, the motherland all brought him heartfelt joy and comfort, but these are not a direct equivalent to happiness. He seemed to pose his own life as an example of philosophical research: What, finally, is philosophy? Moreover, what, finally, should one do to study or pursue philosophy? One might say he dedicated his own thought and theory, and dedicated his own life's practice, to philosophy, the former intentionally, the latter unintentionally. Howsoever it may be, he was a man who gave his thought, his livelihood, and his life to philosophy. Brother Hao and I can be said to have had "A lifelong teacher and friend–this you were to me"; if we meet again, who knows if he'll agree with what I've written here?[9]

---

[8] *Hamlet*, I.iv. The First Folio (and the Dyce edition, e.g.) has "our" for "your."

[9] The line is from a poem by Li Shang-yin (813-858), "Lament for Liu Fen": Liu [1969], 132.

References

Cited writings of Hao Wang:[10]

1974a. *From Mathematics to Philosophy.* London: Routledge and Kegan Paul.

1985a. *Beyond Analytic Philosophy. Doing Justice to What We Know.* Cambridge, Mass.: MIT Press.

1987a. *Reflections on Kurt Gödel.* Cambridge, Mass.: The MIT Press.

1993f. From Kunming to New York (Chinese). *Dushu Monthly* (May), 140–143. English translation in this volume.

1996a. *A Logical Journey: From Gödel to Philosophy.* Cambridge, Mass.: MIT Press.

Other cited writings:

He Zhaowu, 1995. Huainian Wang Hao. *Xi Nan Lian Da xiao you hui jian xun (Southwest Associated Alumni Newsletter)* (October), 48-49. Reprinted in He, *Li shi li xing pi pan lun ji*, pp. 779-782. Beijing: Qing hua da xue chu ban she, 2001.

Liu, James J. Y., 1969. *The Poetry of Li Shang-yin: Ninth-Century Baroque Chinese Poet.* Chicago: University of Chicago Press.

---

[10]For further publication details v. the bibliography in this volume.

Collaborating with Hao Wang on Gödel's Philosophy

Eckehart Köhler

I owe Hao Wang a great debt for encouraging my research on Kurt Gödel, in which we collaborated during the decade-and-a-half before he died.

Of course, I had known *about* Hao for many years as being among the most prominent workers on foundations of mathematics, authoring several important contributions in set theory, computer theory, and philosophy of mathematics. I also particularly respected his didactic efforts to bring mathematical logic to wider audiences than technically trained experts normally face.

Shortly after Gödel's death in 1978, I had begun a project on Kurt Gödel with friends of mine in Vienna, Werner Schimanovich and Peter Weibel. Schimanovich and Weibel invited Wang to contribute to a proposed collection of biographical and logic-oriented articles on Gödel, which ultimately appeared in 2002.[1] At the time, I was living in Astoria, Queens, New York City. I was able to combine work on Gödel with other projects involving the Vienna Circle, since Gödel was a member of the Circle, a fact which was until fairly recently consistently downplayed, not least by Gödel himself.

On a trip to California in 1979, I went to visit Solomon Feferman at Stanford University and told him of our Vienna Gödel project

---

[1]Schimanovich and Weibel initially wanted permission simply to reprint Wang [1981c], which was unique at the time, when very little reliable material was known about the reclusive Gödel, moreover backed by Gödel's direct authority. With the large amount of new material at our disposal, we all agreed to let the request lapse. An important new development was asking Karl Menger to write up a memoir on his relations with Gödel in Vienna and at Notre Dame, which has become a major document on Gödel. It appeared in my English translation as Menger [1994], before our own book, Köhler et al. [2002], whose appearance was after Menger's death.

and of my own interest in understanding and analyzing Gödel's Platonism. Feferman of course was always a constructivist and reserved about Platonism, but rather asked if I had not yet spoken with Hao Wang. So *that* item took a high place on my schedule.

Our Viennese Gödel-project at first was to include reprints of a few of Gödel's major papers. Feferman intended such a project himself, and in the end he was chief editor of Gödel's *Collected Works*. These two projects never really competed, as the Vienna project never intended to translate the German articles, to say nothing about establishing such a stellar cast of editors and commentators as that which Feferman and his co-editors organized. About two years later, in 1981, Feferman inquired with Schimanovich how our project was doing, and the answer was that we had obtained too little funding for any major reprinting project, and since little overlapping activity would take place, Feferman continued organizing his cast of editors and its work schedule.

I probably first met Wang at his office at the Rockefeller University while I was still living in Astoria, in 1979 or early 1980. I had occasionally seen him at logic meetings, or at talks at the Columbia University Philosophy Department, but had not talked with him prior to 1980. I told him about our Vienna project on Gödel. Since Wang had already committed himself to Gödel studies in his later years at Rockefeller University, he was very interested in our work, especially after we (especially Schimanovich) began collecting many very valuable documents on Gödel's biography from his brother Rudolf, who lived as a retired medical doctor in Vienna, from the University of Vienna, and from government offices. I myself subsequently discovered important documents related especially to Gödel's time at the University of Vienna and to his membership in the Vienna Circle (which should more correctly be called the "Schlick Circle"). Collecting material on Rudolf Carnap, the Vienna Circle's most prominent philosopher, from his papers at the University of Pittsburgh Library, I found many interesting documents involving Gödel. In 1985, Schimanovich, Weibel and I interviewed colleagues and acquaintances of Gödel in Princeton and New York for an Austrian TV film on Gödel, broadcast in 1986. After completing the project work on the Vienna Circle in 1984, Schimanovich, the Viennese mathematician Norbert Brunner and I founded the Kurt Gödel Society in Vienna in 1986. We invited Hao Wang to become its founding president, and we were very happy when he agreed to serve in that function. Coincidentally, Wang had been invited as a main speaker to the Wittgenstein Symposium in August, 1986, in Kirchberg am Wechsel. On Wang's visit to Austria, I particularly remember interviews with Wang and Robin Gandy for the TV film on Gödel.

In Vienna, we virtually showered Wang (but also Solomon Feferman and John Dawson) with copies of everything we had, especially correspondence and photographs of Gödel and his relatives in Vienna and Brno. Wang studied certain of these things even more intensely than any of us in Vienna did, so our aid "paid off" in that sense. I will return to these things in sequence below.

The documents I discovered which were perhaps most important to Wang were in connection with research I already started on my project on the Vienna Circle, concentrating mainly on Rudolf Carnap, who in particular was Gödel's first logic teacher. I gave Wang copies of a number of documents I had found in the University of Pittsburgh relating to Carnap and others in Vienna who had relations with Gödel. Wang published many of them for the first time in his [1987a].[2] Wang reciprocated by sending me later in 1986 a package of photocopies of his extensive notes of discussions with Gödel from the period 1969–1975. His notes formed the core of his posthumous *A Logical Journey* [1996a]. They were only ordered by time, and were not otherwise particularly systematic. I profited very much from them, nevertheless, basing on them some of my work on "Gödels Platonismus" [2002].

Curiously, of the biographical material we obtained in Vienna, Wang got most involved with Gödel's decades-long correspondence with his mother in Brno and Vienna, until her death in 1966. We in Vienna found this correspondence frankly rather dreary, because it seemed tedious to us: Gödel mainly wrote his mother about his diet, his health, a minimum about his wife Adele and their domestic situation, a minimum about politics. The only thing we found

---

[2]With one tidbit I got the jump on everybody. I gave a talk [1983] in Salzburg on Carnap and Gödel, and was probably the first to publicly quote the passage from Carnap's diary of August, 1930, relating how Gödel told Carnap and Feigl, among others, at Café Reichsrat (behind the Parliament in Vienna) of having obtained his incompleteness proof, "influenced by Brouwer," as Carnap wrote. Several hundred people in the audience dropped their jaws in awe, as if witnessing a religious apparition.

really interesting was his descriptions of Einstein. Since his mother expressed great interest in Gödel's friendship with the great man, it gave Gödel something to write about. And towards the end of his mother's life, Gödel first wrote about theological and cosmological topics, trying to make his standpoint understandable.

I think Wang was particularly interested in these letters to Gödel's mother because Wang was curious about all aspects of Gödel's personal life. Wang having grown up in China, he was taught the Confucian way of having deep respect for one's parents and for one's teachers, and he approached Gödel with the respect that a Chinese treated an honored teacher. As Wang told me (and presumably others as well), his acculturation was somewhat schizophrenic from early on, as he learned from his parents already as a youngster both the Chinese tradition as well as the modern European-American approach to science, art, and politics brought by the revolution of Sun Yat-sen of 1912. But Wang's approach to Gödel seemed perhaps to bridge the Chinese Confucian tradition with that of western science: Wang chose Gödel as a kind of master, as a great scientist, whom Wang could both respect and learn from, but with whom he could also test his knowledge and vision. Another aspect of Chinese tradition which doubtless motivated Wang, but one going even more deeply than Confucianism, was the Chinese respect for what can be called "roots": not only Gödel's family background was important for Wang, but even the geography and landscape of Gödel's origins were important for Wang. I might mention parenthetically that Wang had always felt a longing to return to China, especially to die there.[3] When I got to know him, he was preparing to accept a very attractive official invitation by Chinese academic officials to teach and do research as long as he wanted in Beijing. During the 1980s, after Deng Xiaoping became Chairman, China was undergoing its opening to the west, allowing hundreds of thousands of students to study in America and Europe, completely reversing the repressions of the "cultural revolution," which had turned Wang away from Maoism and Marxism. But Wang sorrowfully tore up his invitation after the massacre of Tienanmen Square in 1989 and resigned himself to

---

[3]Those familiar with Gustav Mahler's "Lied von der Erde" (songs based on Chinese poems) may remember its main point: the longing of the aging pilgrim for the spot where he was born [*seine Scholle*] to return to, to die on.

remain in the west. His compensation was to intensify his Gödel studies.

Some commentators on Wang have remarked on his way of shying away from definite positions. In the extensive discussions I had with Wang on Gödel's Platonism, and on Carnap's Conventionalism, Wang of course would not take a position. Neither did I, for that matter. Both Wang and I shared a healthy skepticism, always wanting to keep our minds open. But then I resorted to the tool which the Vienna Circle used to decide alleged "pseudo-problems": the Testability Criterion. I searched for evidence which would or could decide between the two positions.[4] So far as I could see, Wang was never too interested in this approach, but preferred his more groping approach. Two things are required for success: **1.** the positions have to be explicated with great precision and care; **2.** evidence from all relevant areas of knowledge must be brought to bear. For both of these endeavors additional hard, sometimes extensive research is required which I thought Wang was too little interested in.

On other topics we found pleasure in discovery and in agreement. One of our favorite topics of discussion was the famous distinction between discovery and invention, well known as a bone of contention between idealists and realists. Here we reached what both of us considered a mature and viable position. The realist (Platonist) says we do not invent mathematical objects, we discover them as they are, we do not make mathematical theories true by convention, we find out through intuition of mathematical "facts." The idealist (Conventionalist) says we "decide" on the acceptability of mathematical theories. Since Wang had been a Marxist for many years, he of course was familiar with Hegel's idealism; and he was familiar with Poincaré's conventionalism; so much so that he correctly pointed out that Carnap's Conventionalism was not the same as Poincaré's. I think Wang is right, and I worked out the details in a forthcoming paper, originally presented at a conference I attended

---

[4]The Testability Criterion (originally the Verifiability Criterion, until it was discovered that many or most interesting theories could not be verified) was of course intended as an Empiricist Criterion of Meaning, as a way to demarcate Science from Metaphysics, so only empirical evidence was permitted. My view is that empirical evidence must be complemented by "reasons" or intuitions, because I recognize two classes of evidence.

with Wang in Oaxaca in 1992. Both Wang and I agreed that the famous distinction is probably illusory. Why? Because conventions are never really free from constraints. If we knew everything there is to know about constraints on conventions, it may be that there is never (or only rarely) any serious alternative to selecting particular rules or principles of action. This is also a view held by objectivists in ethics. Conversely, all of the evidence which the realist calls on in support of his objective decision simply summarize the various constraints which bind the Conventionalist in his choices.

Closely related to this issue is the line between objective *vs.* subjective theories of mathematics. Again perhaps Hegel's Absolute Idealism served as a model for us.[5] In discussing Gödel's distinction between "subjective" and "objective" mathematics, it quickly becomes clear that objective mathematics is simply a limit case of a hierarchy of (more and more objective) subjective theories. Obviously this has immediate relevance to Platonism when interpreted as a limit-case epistemological position.

In August 1986, Wang gave his talk on "Gödel and Wittgenstein" at the 11th Wittgenstein Symposium in Kirchberg am Wechsel, a couple of hours' ride from Vienna (see [1987b]). Wang enjoyed these excursions tremendously, and he took the opportunity to get as much as he could from discussions, from social gatherings, from the dinners and banquets, from the interactions with old acquaintances, with students, and with the townspeople of Kirchberg as well. Although Wang worked much more on Gödel than on Wittgenstein, it was clear that Wittgenstein genuinely interested him—again, from both an intellectual as well as biographical standpoint. I could not myself understand why Wang took Wittgenstein as seriously as he did. Wang did sympathize with Russell's famous criticism of Wittgenstein. In the end, Wang's survey of Wittgenstein in 1986 was actually a little scathing, strongly recommending Bernays's famously rigorous review [1959] of Wittgenstein's *Remarks on the Foundations of Mathematics*.

After the Wittgenstein Symposium of 1986, Wang and I spent some time at museums in Vienna, and we even went to the State Opera to hear Richard Strauss's *Rosenkavalier*. In the late 1980s

---

[5]Of course I was mildly shocked to discover I was sharing anything with that slippery old Hegel. But we give credit where credit is due. Besides, Hegel really got this from Kant, and perhaps from Berkeley and Hume.

and early 1990s, I met regularly with Wang, about once or twice a year, either on his trips to Europe, especially to Vienna, or in New York, usually at Rockefeller University, where Wang several times had me invited as a visitor. Our closest collaboration was on an unpublished manuscript by Gödel, later published as Gödel [*1961/?] in the *Collected Works*. This manuscript especially interested Wang because of the way Gödel contrasts opposing *Weltanschauungen* on foundational questions, giving examples from Kant and Husserl. It had apparently been intended as a lecture at the American Philosophical Society in Philadelphia, which had invited Gödel to become a member in April 1961.[6] The topic was dear to Wang, because for a time he had had a kind of running debate with Gödel on giving Kant preference over Husserl on philosophical foundations, whereas Gödel preferred Husserl. In the end, Gödel ruled it a draw between Kant and Husserl, claiming that Husserl took the best of Kant—a result which appealed to Wang and fascinated him. John Dawson had already shown me a preliminary transcription of the shorthand in Vienna around 1983, and I had already worked on it. John Dawson's wife, Cheryl, had gotten quite skilled in transcribing the Gabelsberger shorthand, being helped by the fact that Gödel did not use many personal abbreviations, sticking rather closely to the standard Gabelsberger he had learned in school in Brno. Wang asked me to help correct the manuscript even more for an English translation he wanted to prepare. A variety of specialized knowledge of Austrian idioms, philosophical and logical terminology was required to get the transcription into shape.

Feferman then invited me to write a commentary on the manuscript for Gödel's *Collected Works*, which I ultimately felt unable to do, feeling too uncertain about Husserl. I also felt that the question of which of Kant or Husserl to prefer on foundations is rather far-fetched. In my view, Gödel was being much too gentle with Kant and Husserl, simply because they are both (undeservedly in my view) so highly regarded. Wang knew my arguments, but had a

---

[6]For the probable background of Gödel's preparation of this text, see Gödel [1995], p. 364. That the text is in Gabelsberger shorthand (and therefore in German) indicates that Gödel had not taken the final steps of preparing a lecture. The conjecture that he never intended to deliver it gets support from the fact that we do not have evidence that Gödel replied to the invitation to give a short lecture.

nostalgic desire to rehearse his old debate with Gödel, culminating in the treatment at §5.3 of his [1996a].

I did not understand what motivated Wang to support Kant's side in his "running debate" on Kant *vs.* Husserl with Gödel. Here I agreed with Gödel's sometimes drastic opinion that important parts of Kant's philosophy, "if read literally," cannot be true. However, I fully agreed with Wang on Husserl. Wang doubted whether Husserl's method of a priori intuition was effective (or even learnable); and Wang could not find instances in Gödel's thinking where Husserl's method played any role.

Wang and I often discussed the old topic of mind *vs.* machine, which he had famously analyzed in his [1974a]. Later on in my [2001], I (re)discovered that von Neumann thought all of logic and mathematics could be physicalized in some sense. But in Wang's library I discovered an interesting book by Benardete [1964]. Its idea is to interpret mathematics from the fictional assumption that we have access to transfinite physical processes. If we assume this, it seemed to me that Gödel's entire motivation for distinguishing between mind and machine completely collapses. Wang did not see the point, and even inquired why I should hold materialism of this sort to be helpful in understanding mind. In the meantime, a bit of philosophical literature has grown up around Benacerraf's concept of "supertasks" in which Benardete plays a role. However, few logicians or philosophers of mathematics have taken it seriously, as far as I am aware; although Hilary Putnam did see that with Benardete's assumption we could "directly verify Platonism."[7] It would not directly "verify" Gödel's Platonism, which was defined epistemically and indirectly through human intuition, but the assumption of supertasks helps to understand the relations between mind and machine; see my [2001].

To Wang, I also pointed out Gödel's insight that "ideal intuition" can be used to test set-theoretical axioms, an approach obviously related to Wang's well-known work on iterative sets. Actually, recalling that Turing Machines were not intended by Turing to be

---

[7]Benacerraf and Putnam ([1964], p. 17, 2d ed. p. 19*f.*) think that number theory could in principle be (Platonistically) verified by assuming supertasks, but not set theory. I do not see why supertasks cannot be "strengthened" sufficiently for this task as well; after all, various theories about "ideal intuition" to verify axioms of set theory *de facto* treat intuition as a supertask-faculty.

physical devices but in fact explications of human or other *mental processes*, it is hard to see why supertasks should not be declared mental directly.

We now entered dangerous turf. Wang had worked for years on computer proofs, growing heartily sick of it, as he confided to me in the 1990s. I presume the reason for Wang's dislike is the simple fact that computer proving quickly gets to be routine drudgework (aptly called "hacking," from "hackwork"), and it is far away from the aesthetically exciting and challenging work of discovering new theorems or even theories which mathematicians prize.

When discussing mathematical discovery and theory formation, I began explaining to Wang the well-known work of Herbert Simon, a major figure in Artificial Intelligence. He and his students had done some of the best work on scientific discovery, using AI methods. (I had at other times tried explaining related concepts of decision theory, statistical weighing of evidence, concepts of probability, in attempting to justify Gödel's famous claim that mathematical knowledge is analogous to empirical knowledge, especially the passage by Gödel where he emphasized that mathematical belief can very well make use of "probabilistic" reasoning, but very little of this interested Wang.) Wang practically exploded. He was disgusted with Simon and did not want to hear a thing about him, considering him an execrable logician. Wang would brook no contradiction, so I changed the topic.[8] Wang's reaction very likely had to do with the work, earlier than his own, of Simon on computer proof.[9]

Without knowing any details of Wang's work in computer proofs, I am sure that he was a logician vastly superior to Simon, and I suppose he got sick of hearing from AI researchers that Simon was allegedly the first to do an AI logic proof. I am also sure that Simon would agree with Wang that the proof he presented at the famous 1956 AI-conference was utterly trivial and without much inherent merit. Nevertheless, Simon's overall scope of knowledge, and his unfailingly wise advice on many topics of scientific research, on social science and economics, and most especially on the concepts of rationality and intelligence, make him one of the most interesting

[8]Oddly enough, Simon [1991], ch. 22, spent quite a lot of time in China and had learned Chinese.
[9]See Newell, Shaw, and Simon [1957].

writers of all in philosophy of science.

Wang, in sharp contrast to Simon, did very little reading outside of the areas that immediately interested him, and I noticed that his library was rather smaller than mine. Perhaps I am interested in too many different things and get too sidetracked; e.g. Wang advised me not to read newspapers so much. This brought me to my only serious criticism of him: that he was not paying enough attention to "what we know." I refer of course to Wang's *Beyond Analytic Philosophy: Doing Justice to What We Know* [1985a], of which I have an extremely high opinion. There is absolutely no better book on analytic philosophy available, and when the *New York Times* asked me to summarize Wang's philosophy when he died in 1995, I referred above all to this book. (I assumed his technical work in logic would not interest a lay public.) The book's pinnacle is already its introduction, whose theme Wang returns to in its last chapter. Its point is that the philosopher should respect "what we know" and not try to philosophize independently from it, nor should he try to overturn it all in a radical putsch. Wang's critical review of the mainstream of analytic philosophy from Russell through Carnap and Quine is absolutely commanding, and his main point of criticism is that they, especially the latter two, did not take our knowledge of higher mathematics seriously enough, trying instead to regiment it by use of preconceived philosophical schemes. Of course, Wittgenstein's work was an extreme case of this syndrome, but Wang lets him off lightly.

A few days after the clash on Simon, while musing on Gödel's notion of intuition, I mentioned how close it seemed to what AI people (following Simon) now call heuristics; it is connected with "rule-based" programming. Wang again rebuffed me, calling this too "scientistic."[10] I slyly hinted that his refusal to consider the idea violated his own injunction "to do justice to what we know." Wang

---

[10]Yes, I *was* guilty of being "scientistic"; but so was Leibniz, so was Gödel. It means to look for the best and latest scientific authority, and not shying from technical difficulty. I quote from *Webster's Third New International Dictionary* (known as *Webster's Unabridged*):

scientism ... **2** : a thesis that the methods of the natural sciences should be used in all areas of investigation including philosophy, the humanities, and the social sciences : a belief that only such methods can fruitfully be used in the pursuit of knowledge.

scientistic ... **2** : of, relating to, or characterized by scientism.

of course immediately got the point and did not reply. On my part, I never tried to force my own positions on him, and if he did not take up an idea, I dropped it. We were independent individuals, and we respected each other's opinions and independence. I concluded that Wang was not being sufficiently problem-oriented, looking for constructive, if necessary original solutions to old difficulties, such as understanding Platonism.

Whereas Wang's [1985a] was somewhat circumspect in his respectful critique of Quine, Wang's contribution [1986b] to Quine's "Schilpp volume" was a little different, even though it continued the line of attack concerning doing justice to what we know. Particularly on pp. 633-34, Wang notes a number of major developments in logic and foundations of mathematics that are not reflected in Quine's research and ends by remarking that "for the working logicians, much of Quine's work is thought to be off the mainstream" (*ibid.*, p. 635).

One thing Wang and I heartily agreed on was our disappointment with Quine's philosophy, which Wang jocularly called Quine's (logical?) "negativism" ([1985a], §16.6). Whereas Quine contributed positively to set theory, his general philosophy clearly was dominated by negative positions: rejection of the analytic/synthetic dichotomy, and his attack on semantics, peaking in the claim of the indeterminacy of translation, were very well known. Quine also strongly rejected intensional logics, to the study of which even Gödel contributed. There was also a kind of blitheness of Quine's mentality which Wang and I disliked. Wang and I agreed that Quine was terrifically "banal." This of course is not really a serious personal defect for any normal man. In fact, Quine was one of the best-mannered men I ever met (speaking now for myself), but for a philosopher, banality is deadly.

Neither Wang nor I want to make an "enemy" out of Quine. Both Wang and I loved to read Quine, and have learned many things from him. I suppose we were just disappointed that Quine cannot play the role of a major thinker for us. Gödel is such a major thinker for us. Gödel had his quirks, but he was as "unbanal" as one can imagine. The difference is that Quine was much too dominated by his prejudices, and we want freedom from prejudice. Gödel was as unprejudiced as a thinker can be.[11] On Kant and

---

[11]Not entirely true. Gödel was indeed *sometimes* prejudiced. He made a

66

Wittgenstein, Wang and I disagreed, however. I found that Kant's reputation has been greatly exaggerated. On Wittgenstein, I asked Wang, if he thought it so important to study and understand Gödel, how he could possibly take someone as confused and ill-informed as Wittgenstein so seriously. Of course, Wang knew the criticisms of Kant and Wittgenstein, but he had no answer.

Finally Wang [1987a] appeared, which he inscribed for me. I was extremely impressed by its deep devotion and scholarship, by its struggle to understand and not to avoid complexities. I immediately read it and made extensive notes. This volume did not yet make use of the notes of conversations with Gödel which Wang had let me copy (which I had meanwhile already been making keen use of). Instead, Wang made use of the biographical material he had collected, some of which came from Schimanovich and myself. I think Wang finally overcame his legendary unhappiness working on Gödel.[12] I think that the book provides a model of tying philosophical disquisition together with biography and contemporary history. What impressed me the most was Wang's treatment of religion and theology. Most people know that Chinese thought is non-theological, not having been (much) influenced by Persian dualistic theology or by Judeo-Christian-Islamic monotheism. Neither Confucianism nor (orthodox) Buddhism makes much use of theology. When Wang first began his discussions with Gödel, they agreed *not* to discuss

serious mistake in identifying the machine concept with Turing Machines, which Kreisel [1974], [1980] has shown to be mistaken. Gödel also made a mistake in tying Platonism together with Plato's illusion that ideas are abstract and hence constitute a realm distinct from empirical reality, which was concrete (critics of Platonism call this the "Verdoppelungshypothese"). I have shown in Köhler [2002] that abstract entities belong to the empirical world as much as concrete objects do, and that Platonism therefore cannot assume their separation at all. Whereas this latter issue is very laden with philosophical-historical freight, and Gödel could perhaps be "excused" for not seeing the point, Kreisel's correction of Gödel's mistake is absolutely convincing to anyone familiar with the history of mechanics, as Gödel was, since he had studied physics, and Kreisel [1980] relates how Gödel relented the minute it was stated; nevertheless, Gödel later on reverted back to his prejudiced view—because he was so "wedded" to the idea that mind can do more than machines (= matter); whereas if machines were suddenly nonfinitary (with reference to Hilbert's sense of finitarily computable), as e.g. the solar system in fact was with Newton, this is no longer so clear.

[12]Quine ([1985], p. 306) represents Wang as persistently unhappy during the earlier stage of his career.

theology, despite Gödel's intense interest (maybe *because* of it?). In contrast to Wang, I was familiar with theology, having been raised a Quaker (Society of Friends). I had later also learned to understand theology in connection with metaphysics and logic through Heinrich Scholz [1921] and [1961], who held that modern logic is the "epochal" form of metaphysics (and therefore theology) for the modern world—a very Leibnizian view.[13] I was therefore worried that Wang might give too little space to discussing theology, which of course would be a serious omission in writing on Gödel.

Once, on a walk along the East River near Rockefeller University, Wang asked me, in his inimitable, leisurely way, why heaven is so important in western religions. The answer came in a flash: heaven comes from Persian dualism (which influenced Pythagoras and Plato very much, and is really the origin of Platonism). The Jews did not have paradise in an other-worldly heaven, but rather here on earth—a serious heresy for Christians. In accepting Zoroaster's identification of paradise with the other-worldly Persian heaven, Christians bound up a utilitarian idea, making it the soteriological-eschatalogical goal of life. This is of course far from standard utilitarianism, which renounces other-worldly goals not connected to earthly pleasures (not to mention the idea of an infinite utility). To make a long story short, Zoroaster told his followers that, if we do on earth what God (i.e. Zoroaster's good God, Ahura-Mazda) wants, which is the Good, then at the end of history, on judgment day, God will admit us to heaven; which to attain Pascal later interpreted to be the infinite payoff for placing a bet that God exists. This was already very decision-theoretical, and I feared that this might put Wang off, but the topic was so interesting to him that he ignored my over-technical references to utility.

I was surprised and very impressed by Wang's guileless question, since he had been after all an atheistic adherent of Maoist communism, and had previously brusquely refused to talk about the opiate of the people. Now he simply dispensed with his old attitude out of curiosity over Gödel's beliefs. And Wang ([1987a], ch. 8, [1996a], ch. 3) really did his homework, displaying amazing sophistication and subtlety about theology. Of course, Wang

---

[13]Scholz came to this view when he discovered *Principia Mathematica* in 1921, but Russell [1900] already wrote how tightly connected Leibniz's logic and metaphysics were.

was not really so naïve after all, since he had had a good training in philosophy, including Leibniz and Hegel, for both of whom theology played the central role in metaphysics. There is, in my opinion, another, deeper factor explaining why it was possible for Wang to so readily discuss theology, despite disclaimers that it still had no effect on his "life in any way" ([1987a], p. 217). This is that the mathematics-saturated world-view of Pythagoras is responsible for much of Christian theology, since Saint Augustine and many of the patristic theologians were thoroughly acquainted with neo-Platonism, which was dominated by Pythagoreanism. Anyone like Wang who is acquainted with Cantorian theories of the transfinite ordinals is therefore enabled to easily understand the workings of an infinite mind.

In the 1990s, I saw somewhat less of Wang than before, as I had taken on a teaching job in Vienna and had less time. The Gödel project of Schimanovich, Weibel and myself became endlessly complicated with changing personnel, shifting contents, and expansion to two volumes. My three contributions on "Gödel und der Wiener Kreis," "Gödels Jahre in Princeton," and "Gödels Platonismus" had been completed by 1990, but the volumes appeared distressingly late in 2002. My last visit with Wang in 1994 I remember with great poignancy, as he had surprisingly come to the gate of Rockefeller University to welcome me, despite the fact that he was weakened by the cancer that he was to die from the next year. I went with Wang first to his office, then to his home, where he lived with his third wife—his happiest and most successful marriage. (His wife was a German woman born in Prague who had become a dance performer in the US, so we had much to talk about, since my own parents married and lived in Prague in the 1930s.) I was overjoyed that he seemed to be in as good health when I saw him, apparently enjoying the wonders of medicine to a large extent brought about by research right there at Rockefeller University itself. But I still was saddened by the thought that it would be the last I would see of him—as it was. When he died in 1995, *The New York Times* prepared an obituary, and Marie Grossi, Wang's loyal and understanding secretary, told them to call me in Vienna for help about his philosophy. I was at first nonplussed and unprepared, because my discussions with Wang had never had "his philosophy" as their topic, but always Gödel's, or the Vienna Circle's, or Russell's, but

I was able to prepare a sketch of Wang's major contributions to set theory, computer proofs, work on the mind–machine relationship, his critique of analytic philosophy, and of course his greatest achievement in philosophy, his devoted analysis of Gödel's thought.

To summarize the narration of my collaboration with Hao Wang in studying Gödel's philosophy, I learned much and was as a result able to improve my understanding very much, especially about topics related to the contrast between subjective and objective views, and about relations between mind and matter (machine). Whereas I achieved almost complete harmony with Wang about subjectivity/objectivity, Wang saw little point in my ideas about mind/matter, although I have not let up in my attention. I had hoped I would eventually persuade Wang of the importance of the topic, but there was too little time left. I liked Wang's method of closely connecting the theoretical interests of Gödel and other thinkers with their biographies and general views about life—striving truly for a holistic view of life. For the work of great men is always tied to their ways of life and to their character, and one can learn as much from the latter as from the former.

My feeling was often that Wang took on with me the role of an avuncular mentor, which suited me well, especially since I had little contact with my own father after he divorced my mother. Whether I fulfill any hopes Wang had for my work the future will tell, but I am certain of its value, for I had never before felt that I was contributing important new solutions to classic problems of philosophy as I had after beginning my research on Gödel (in conjunction with studying Carnap, I may add). What I cherished the most about Wang was a kind of lust for life that he had, which he had despite his earlier legendary unhappinesses. I was especially pleased by his lust for woman's companionship, beginning his third marriage at around 70 years of age.

70

References

Cited writings of Hao Wang:

1974a. *From Mathematics to Philosophy*. London: Routledge and Kegan Paul.

1981a. *Popular Lectures on Mathematical Logic*. Beijing: Science Press. Also New York: Van Nostrand Reinhold. Reprinted with a Postscript, New York: Dover Publications, 1993.

1981c. Some facts about Kurt Gödel. *The Journal of Symbolic Logic* 46, 653-659.

1985a. *Beyond Analytic Philosophy. Doing Justice to What We Know*. Cambridge, Mass.: MIT Press.

1986b. Quine's logical ideas in historical perspective. In Lewis Edwin Hahn and Paul Arthur Schilpp (eds.), *The Philosophy of W. V. Quine*, pp. 623-643. The Library of Living Philosophers, volume 18. Chicago and La Salle, Ill.: Open Court.

1987a. *Reflections on Kurt Gödel*. Cambridge, Mass.: The MIT Press. Paperback edition, 1990. Japanese translation: Sangyo Tosho, 1988. French translation: Paris: Armand Colin, 1991. Spanish translation: Alianza, 1992. Korean translation: Seoul: Minumsa Publishing, 1997. Chinese translation, with a new preface: Shanghai: Shanghai Translation Publishing House, 1997.

1987b. Gödel and Wittgenstein. In Paul Weingartner and Gerhard Schurz (eds.), *Logic, Philosophy of Science and Epistemology*, pp. 83-90. Proceedings of the 11th International Wittgenstein Symposium, Kirchberg am Wechsel, Austria, 4-13 August, 1986. Vienna: Hölder-Pichler-Tempsky.

1996a. *A Logical Journey. From Gödel to Philosophy*. Cambridge, Mass.: MIT Press.

Other cited writings:

Benacerraf, Paul, and Hilary Putnam (eds.), 1964. *Philosophy of Mathematics: Selected Readings*. Englewood Cliffs, N. J.: Prentice-

Hall. 2d ed., Cambridge University Press, 1983.

Benardete, José, 1964. *Infinity. An Essay in Metaphysics.* Oxford: Clarendon Press.

Bernays, Paul, 1959. Betrachtungen zu Ludwig Wittgensteins "Bemerkungen über die Grundlagen der Mathematik." *Ratio* (German edition) 1959, Heft 1, 1–18; translation in English edition of *Ratio* 2, 2-22, reprinted in Benacerraf and Putnam [1964] (not in 2d ed.).

Gödel, Kurt, *1961/?. The modern development of the foundations of mathematics in the light of philosophy. In Gödel [1995], pp. 374-387.

———, 1986. *Collected Works*, Volume I: *Publications 1929–1936*. Edited by Solomon Feferman, John W. Dawson, Jr., Stephen C. Kleene, Gregory H. Moore, Robert M. Solovay, and Jean van Heijenoort. New York and Oxford: Oxford University Press.

———, 1990. *Collected Works*, Volume II: *Publications 1938–1974*. Edited by Solomon Feferman, John W. Dawson, Jr., Stephen C. Kleene, Gregory H. Moore, Robert M. Solovay, and Jean van Heijenoort. New York and Oxford: Oxford University Press.

———, 1995. *Collected Works*, Volume III: *Unpublished Essays and Lectures.* Edited by Solomon Feferman, John W. Dawson, Jr., Warren Goldfarb, Charles Parsons, and Robert M. Solovay. New York and Oxford: Oxford University Press.

Köhler, Eckehart, 1983. Gödel and the Vienna Circle: Platonism vs. formalism. *Abstracts of the Seventh International Congress of Logic, Methodology and Philosophy of Science, Salzburg, July 11-16,* 1983, vol. 6, pp. 106–108.

———, 2001. Why von Neumann rejected Carnap's dualism of information concepts. In Miklós Rédei and Michael Stöltzner (eds.), *John von Neumann and the Foundations of Quantum Physics.* Vienna Circle Institute Yearbook 8. Dordrecht: Kluwer.

———, 2002. Gödels Platonismus. In Köhler *et al.* [2002], II, 341-386.

———, forthcoming. Gödel's Platonism vs. Carnap's conventionalism.

72

————, forthcoming. *Intuition Regained*, manuscript for a book.

Köhler, Eckehart, Peter Weibel, Michael Stöltzner, Bernd Buldt, Carsten Klein, and Werner DePauli-Schimanovich-Göttig (eds.), 2002. *Kurt Gödel. Wahrheit und Beweisbarkeit*. Band I, *Dokumente und historische Analysen*. Band II, *Kompendium zum Werk*.[14] Vienna: Österreichischer Bundesverlag – Hölder-Pichler-Tempsky.

Kreisel, Georg, 1974. A notion of mechanistic theory. *Synthese* 29, 11–26.

————, 1980. Kurt Gödel, 28 April 1906–14 January 1978. *Biographical Memoirs of Fellows of the Royal Society* 26, 148–224. Corrections, *ibid.* 27, 697, and 28, 718.

Menger, Karl, 1994. Memories of Kurt Gödel. In Menger, *Reminiscences of the Vienna Circle and the Mathematical Colloquium*, edited by Louise Golland, Brian McGuinness, and Abe Sklar. Dordrecht: Kluwer. (Translation of a German text; the original appears in Köhler *et al.* [2002], I, 63-81.)

Newell, A., J. C. Shaw, and H. A. Simon, 1957. Empirical explorations of the logic theory machine: A case study in heuristics. *Proceedings of the Western Joint Computer Conference*, pp. 218-230. New York: Institute of Radio Engineers.

Quine, W. V., 1985. *The Time of My Life: An Autobiography*. Cambridge, Mass.: MIT Press.

Russell, Bertrand, 1900. *A Critical Exposition of the Philosophy of Leibniz*. Cambridge University Press. Later reprints London: Allen & Unwin.

Scholz, Heinrich, 1921. *Religionsphilosophie*. Berlin: Reuther & Reichard.

————, 1961. *Mathesis Universalis: Abhandlungen zur Philosophie als strenge Wissenschaft*. Edited by Hans Hermes, Friedrich Kambartel, and Joachim Ritter. Basel and Stuttgart: Benno Schwabe.

Simon, Herbert, 1991. *Models of My Life*. New York: Basic Books. Paperback edition, Cambridge, Mass.: MIT Press, 1996.

---

[14]In volume II the editors are listed in a different order, with Buldt first.

Hao Wang's Contributions to Mechanized Deduction and to the Entscheidungsproblem

Martin Davis

1. The Predicate Calculus and the Entscheidungsproblem

Predicate calculus (also known as First Order Logic) uses symbols for propositional connectives and quantifiers, typically:

$$\neg \quad \vee \quad \wedge \quad \supset \quad \forall \quad \exists,$$

the symbol "$=$" for identity, variables, e.g.,

$$x \ y \ z \ x_1 \ y_1 \ z_1 \ x_2 \ \ldots,$$

punctuation marks

$$, \quad (\quad)$$

and symbols for constants, functions, and relations.

Here is an example of a sentence of predicate calculus:

$$(\forall x)(\forall z)(\neg(x = 0) \supset (\exists y)(x \cdot y = z))$$

A sentence is called *valid* if it is true no matter how the symbols for constants, functions, and relations are interpreted. According to Gödel's completeness theorem of 1930, the usual rules of proof yield all valid sentences and only those.

The **Entscheidungsproblem** called for an algorithm for testing a given sentence for validity. Church and Turing each proved that this problem is unsolvable, that no such algorithm exists. Hilbert had referred to the Entscheidungsproblem as "the fundamental problem of mathematical logic." Its importance stems from the following fact:

If $T_1 T_2 \ldots T_n$ are axioms for some part of mathematics, then T is provable from those axioms if and only if the sentence $((T_1 \wedge T_2 \wedge \ldots \wedge T_n) \supset T)$ is **valid**.

While a sentence is *valid* if it is true no matter how the symbols for constants, functions, and relations are interpreted, it is said to be *satisfiable* if it is true for at least one way that the symbols for constants, functions, and relations can be interpreted. Evidently:

A sentence Q is valid if and only if $\neg Q$ is not satisfiable.

Thus, the Entscheidungsproblem may equivalently be stated as seeking an algorithm to test a given sentence for satisfiability.

Simple algorithms exist to transform any sentence into a logically equivalent one in which a string of "quantifers" ($\forall$ and $\exists$) precede a "quantifier-free" part. Such sentences are said to be in *prenex form*. Thus the following is a prenex form of the sentence exhibited above:

$$(\forall x)(\forall z)(\exists y)(\neg(x = 0) \supset (x \cdot y = z))$$

Prenex sentences are conveniently classified according to their quantificational prefixes. So, e.g., the *class* $\forall\forall\forall\exists\forall$ means the class of all sentences of the form:

$$(\forall x)(\forall y)(\forall z)(\exists u)(\forall v)Q$$

where Q contains no quantifiers. Such a class is called *decidable* if there is an algorithm that can test any given member of the class to determine whether it is satisfiable. It is called a *reduction class* if there is an algorithm that will transform an arbitrary given sentence Z into a sentence Z^* belonging to the class which is satisfiable if and only if Z is.

In what follows, we restrict ourselves to sentences with no occurrences of $=$ or of function or constant symbols. It has turned out that the presence or absence of $=$ in the formulation of predicate calculus makes a difference in the status of certain of the prefix classes, although this fact had not always been clear.

Evidently (because of the unsolvability of the Entscheidungsproblem), no class can be both decidable and a reduction class. It has been shown that the classes

$$\exists \cdots \exists \forall \cdots \forall$$

and

$$\exists \cdots \exists \forall \forall \exists \cdots \exists$$

are decidable, while the classes

$$\exists \cdots \exists \forall \forall \forall \exists \cdots \exists$$

$$\forall \forall \exists$$

and

$$\forall \exists \forall$$

are reduction classes.

   This gives a complete classification of sentences with respect to their quantificational prefix. The last class mentioned, $\forall \exists \forall$, remained open for a considerable period. The proof that it is indeed a reduction class was given in a crucial paper by Kahr, Moore, and Hao Wang [1962b]. Hao Wang's role in achieving this important result is evident not only in that he is one of the authors of the key article, but also because the proof makes use of a kind of "domino" problem that had been studied by Hao Wang [1961a] earlier. Techniques that had been used previously by Büchi [Büchi, 1982] in his work on the Entscheidungsproblem seem to have also played a role. For an up-to-date study of the Entscheidungsproblem and its ramifications, see [Börger et al., 1997].

2. Contributions to Mechanized Deduction

The problem of programming computers to carry out deductions, such as those that occur in mathematical proofs, is particularly attractive to researchers investigating the extent to which human intelligence can be simulated by machines. This is because, although this arena is one in which insight and creativity are seen to play an important role, the extent to which success has been achieved can be judged in a relatively objective manner, as compared, for example, with programs intended to write poetry or to compose music. Because of the fact, mentioned above, that all mathematical proofs can be thought of as taking place within the predicate calculus, it is natural to make first order logic the focus of this work. However, the unsolvability of the Entscheidungsproblem is a serious obstacle.

Nevertheless, there do exist "proof procedures" for the predicate calculus.

A **proof procedure** *for predicate calculus is an algorithm that when supplied with a valid sentence as input will eventually furnish a "proof" for that sentence. (If the input is not valid, the algorithm may never terminate.)*

Now there are proof procedures that are quite hopeless from the point of view of computer implementations. For example, one could simply use one of the standard formulations of the rules of proof of predicate calculus to systematically generate all valid sentences in succession. Then given a particular sentence to test, one could simply compare it to each valid sentence as it is generated, terminating the procedure if and when the sentence generated is identical with the one being tested. However work during the 1920s and 1930s by Skolem, Herbrand, and Gentzen formed the basis for far more practical proof procedures, and during the late 1950s and early 1960s, a number of logicians, including Hao Wang, proposed and implemented proof procedures based on their ideas. For some of this history, see [Davis, 2001].

It may be of interest to note that Hao's desire to achieve some facility as a computer programmer was originally motivated by his intention to return to China and to participate in the work of the Communist regime. He had hoped that this skill would be more useful than his expertise as a logician. Like so many others, he eventually became disillusioned with the Maoist order. I remember particularly his telling me of his astonishment at a letter severely criticizing him that he received from his father, a secondary school teacher in China, during the Cultural Revolution. He understood with considerable distress that his father would have written such a letter only under great pressure.

The procedures devised and implemented by Hao Wang were based on Gentzen's sequent calculus. He was able to arrange matters so that for given sentences belonging to certain decidable classes, his procedure would always eventually halt (whether or not the given sentence was valid). He reported his work in a number of influential papers: [1960a], [1960b], [1961a], [1963e]. (The methods developed in [1961a] in particular played a crucial part in the later work on the $\forall\exists\forall$ case of the Entscheidungsproblem.) Wang was able to report a seemingly spectacular achievement: a computer program that proved all formulas of predicate calculus that had been proved in Whitehead and Russell's *Principia Mathematica*—over 350 sentences proved in a few minutes. The secret of this success was that, as it turned out, all these valid *Principia* sentences fell into the $\forall \cdots \forall \exists \cdots \exists$ class, and so in the decidable $\exists \cdots \exists \forall \cdots \forall$ class for satisfiability.

To encourage research in the field of mechanized deduction, a special award was introduced for achievements that constituted "milestones" in this field. The first of these Milestone Prizes was awarded to Hao Wang in 1983. The accompanying citation, written by Martin Davis, David Luckham, and John McCarthy read as follows:

> The first "milestone" prize for research in automatic theorem-proving is hereby awarded to Professor Hao Wang of Rockefeller University for his fundamental contributions to the founding of the field. Among these, the following may be listed:
>
> 1. He emphasized that what was at issue was the development of a new intellectual endeavor ... which would lean on mathematical logic much as numerical analysis leans on mathematical analysis.
>
> 2. He insisted on the fundamental role of predicate calculus and of the "cut-free" formalisms of Herbrand and Gentzen.
>
> 3. He implemented a proof-procedure which efficiently proved all of the over 350 theorems of Russell and Whitehead's *Principia Mathematica* which are part of the predicate calculus with equality.
>
> 4. He was the first to emphasize the importance of algorithms which "eliminate in advance useless terms" in a Herbrand expansion.
>
> 5. He provided a well thought out list of theorems of the predicate calculus which could serve as challenge problems for helping to judge the effectiveness of new theorem-proving programs.

References

Cited writings of Hao Wang:

1960a. Toward mechanical mathematics. *IBM Journal of Research and Development* 4, 2-22. Reprinted in Siekmann and Wrightson [1983].

1960b. Proving theorems by pattern recognition, part I. *Proceedings of the Association for Computing Machinery* 3, 220-234. Reprinted in Siekmann and Wrightson [1983].

1961a. Proving theorems by pattern recognition, part II. *Bell System Technical Journal* 40, 1-41.

1962b. (With A. S. Kahr and Edward F. Moore.) Entscheidungsproblem reduced to the $\forall\exists\forall$ case. *Proceedings of the National Academy of Sciences U. S. A.* 48, 365-377.

1963e. The mechanization of mathematical arguments. In N. C. Metropolis, A. H. Taub, John Todd, and C. B. Tompkins (eds.), *Experimental Arithmetic, High Speed Computing and Mathematics*, pp. 31-40. Proceedings of Symposia in Applied Mathematics, vol. 15. Providence: American Mathematical Society.

Other cited writings:

Börger, Egon, Erich Grädel, and Yuri Gurevich, 1997. *The Classical Decision Problem.* Berlin and Heidelberg: Springer-Verlag.

Büchi, J. Richard, 1982. Turing machines and the Entscheidungsproblem. *Mathematische Annalen* 148, 201-213.

Davis, Martin, 2001. The early history of automated deduction. In Alan Robinson and Andrei Voronkov (eds.), *Handbook of Automated Reasoning*, pp. 5-15. Amsterdam: Elsevier.

Siekmann, Jörg, and Graham Wrightson (eds.), 1983. *Automation of Reasoning: Classical Papers on Computational Logic 1, 1957-1966.* Berlin and Heidelberg: Springer-Verlag.

Sets and concepts, on the basis of discussions with Gödel[1]

Hao Wang

Edited with an introduction and notes by Charles Parsons

Introduction

§1. Hao Wang had two series of extended conversations with Kurt Gödel, one important theme of which was the concept of set and the foundations of set theory. The first series, in 1971-72, occurred when Wang was finishing his book *From Mathematics to Philosophy* ([1974a], cited hereafter as FMP). Chapter VI of that work, on the concept of set, contains some clear reports of Gödel's views and was very likely more globally influenced by him.[2] Wang tells us that in the first part of the second series of conversations in the fall of 1975 they took as an "initial frame of reference" the revision of the

---

[1]This text is what Wang refers to as fragment Q ([1996a] (cited hereafter as LJ), p. 149), the latest of four texts that Wang wrote in the fall of 1975 to accompany his discussions with Gödel of the concept of set and related concepts. The source text is documents no. 013167 and no. 013169 of the Kurt Gödel papers. These documents and the quotations from others given here are published courtesy of the Kurt Gödel Papers, The Shelby White and Leon Levy Archives Center, Institute for Advanced Study, Princeton, NJ, USA, on deposit at Princeton University. I am indebted to Montgomery Link for helpful comments and to Robin Rollinger and especially Cheryl Dawson for transcriptions of Gödel's shorthand.

The title on the manuscript is "Objectivism of sets and concepts." As noted in §3 of the introduction, that was to be the title of a three-part paper, whose second and third parts were apparently not written. We use a title suggested by Gödel in remark 9.3.1 of LJ.

Bracketed numbers [2], [3], etc. have been inserted into the text to mark page breaks in the typescript. [n] signifies the beginning of page n. All notes are due to the editor. The italics in the section titles were also introduced by the editor.

[2]See FMP, p. x, on passages that Gödel acknowledged as stating his views. On his role in the final stages of the work on the book, see §3 of the introductory note to the Gödel-Wang correspondence, Gödel [2003a], pp. 385-89.

talk "Large sets" that he had given the previous summer at a symposium on the concept of set at the Fifth International Congress of Logic, Methodology, and Philosophy of Science in London, Ontario.[3] From early November to December 1975, Wang wrote what he describes as four "fragments" on sets, concepts, and realism about them (called objectivism in these texts), on all of which Gödel commented. The conversations during this period were with one exception by telephone, since Gödel had largely ceased to go to his office at the Institute for Advanced Study. Wang called these fragments M, N, C, and Q.[4]

Some of the content of these fragments made their way into Wang's publications, either [1977a] itself or as remarks attributed to Gödel in LJ. They duplicate each other to a certain extent, and only fragment Q is close to being an essay. Since it incorporates much of the content of the earlier fragments, only it is published here, in spite of Wang's testimony that Gödel liked fragment M better.

In fact, Gödel was not happy about Wang's exposition, at least as it was before the handwritten emendations that are incorporated into the text published below.

In what follows I will give some indication of the content of the four texts and their relation to Wang's publications. I will then discuss more fully the content of fragment Q and some substantive issues it raises.

§2. *Fragments M, C, and N.* The copy of fragment M in Gödel's papers bears the date November 17, 1975; it is evidently the earliest of the four texts.[5] It does not properly have a title but begins with the section heading "1. Sets and concepts." However, in contrast to Q the typescript says little about concepts. The first seven pages are quite close to the opening of §1 of [1977a], and nearly all of

---

[3]LJ, p. 148. Wang's paper appeared as Wang [1977a]. I was the co-symposiast; see my [1977]. Since a part of it was devoted to criticizing Wang's justifications of axioms of set theory in FMP, a little more will be said about it below.

[4]LJ, p. 149. Further statements about Wang's testimony concerning these fragments and discussions of them with Gödel refer to pp. 148-49 of that work.

[5]It is no. 013161 of the Gödel papers. I note that we rely on this collection for all our documents. Light on some issues might be shed by inspecting copies in Wang's papers in the Rockefeller University Archives. However, fifteen years after his death his papers have still not been made available.

the rest surfaces either in that paper or in Q. A brief argument for realism (pp. 3-4) is omitted in [1977a], but a fuller version appears in §1.7 of Q. The text makes remarks about the thesis of FMP that a multitude (or plurality) is a set if it has an "intuitive range of variability." This point is also addressed in Q and in [1977a], pp. 312-13, and we will comment on it later.

Fragment C is focused on concepts and can be viewed as a preliminary version of part of fragment Q. There is nothing of note that did not find its way into that text.[6]

Fragments M and C were the basis of the one discussion in person that Wang had with Gödel during this period, on December 9, 1975. Wang soon completed fragment Q; one might guess that he had been working on it before.[7] Since M is focused on sets and set theory while C is focused more on concepts, one might view the writing of Q as an attempt to synthesize these two aspects of Gödel's views. Gödel's preference for M may in part reflect the fact that he was more confident in his views about sets than in his views about concepts.

In fact, Gödel's reaction to the second installment of fragment Q was rather negative. On the envelope he wrote, "no clear argument" and below it "chaos of various assertions," and a little further to the right, "disorganised."[8] On the top of the first page he wrote

---

[6]This is document no. 013164. Wang writes that this text was the basis for a lecture to the Association for Symbolic Logic in March 1979. The lecture was entitled "Report on some of Gödel's philosophical views" and was given on 24 March in a symposium on Gödel at a meeting of the Association in San Diego. (See *The Journal of Symbolic Logic* 46 (1981), p. 199.) I have no further information about the content of the lecture. Fragment N is document no. 013165.

[7]The first page has the handwritten date December 11, 1975. In all probability this is the date of the typescript of pp. 1-9, since the covering letter with which Wang sent that to Gödel is dated December 12 (Gödel [2003a], p. 414). The remainder was sent to him on December 15. (Gödel [2003a], p. 415 note a, is in error in stating that it was sent on December 19; the postmark on the envelope (document no. 013168) is very clear.) Gödel's annotations are certainly later, and Wang's handwritten changes presumably reflect discussion with Gödel and are likely to be still later. Wang says that Gödel commented on it extensively on 4 January 1976.

[8]I am grateful to Mark van Atten for calling this envelope to my attention and supplying a copy and to Robin Rollinger for transcribing the shorthand. It should be said that crotchety responses to writings sent to him were not unusual for Gödel at this time.

"[He] intersperse[s] inessential things & distract[s] the attention on the other hand the main points of argument often concealed." The same place contains a shorthand remark that can be translated, "He should have [included] literally everything in the original version, only a few small [things] from printed [things] and additions."[9]

That raises the question, What could Gödel have meant by "the original version"? Since he wrote "second version" on the envelope, he must have had in mind some first version. Furthermore, Wang wrote in the covering letter of December 15 sent with the second installment, "I do feel the version is better than the 'main version' which you liked before. I hope you agree with me."[10]

What is most likely is that both refer to fragment M, or possibly to M and C together. No other document in Gödel's papers is a plausible candidate. However, we cannot be entirely certain until it is possible to see what has been preserved in Wang's own papers.

The reader should judge how much justice there is in these remarks of Gödel; in particular, we do not know how considered they were. We can assume that most of the handwritten changes to the typescript arose from discussion with Gödel, but they would suggest that, instead of sending Wang back to the drawing board, Gödel encouraged more piecemeal revision.

Thus far we have said nothing about fragment N, entitled "Notes on sets and concepts." It is an outlier among these documents, since Wang introduced into it quite other philosophical ideas of Gödel and also uses quotations from Engels and Marx. Wang says of it, "I tried to combine some of Gödel's ideas with my own current interest in Marxism." It was evidently in Gödel's hands before the meeting of December 9. Gödel's annotations indicate that he reacted negatively to some statements in it. He probably found the remarks about Plato and Aristotle too favorable to Aristotle's criticisms of Plato. Gödel's annotations of the following paragraph are revealing:

> We should look more closely at the place of the objectivistic
> picture in Gödel's mathematical practice. We could then see

---

[9] For the German and a version of these comments closer to the raw texts, see the textual notes.

[10] Together with the envelope, this letter is document no. 013168 in the Gödel papers. It is not published in Gödel [2003a]. Thanks to Mark van Atten for reminding me of this remark.

that the existence of mathematical objects is, among other things, a useful heuristic picture. In fact, his mathematical work has influenced people to entertain the same sort of picture he has in mind. And he wants to speak clearly about the picture in order to give, for example, a satisfactory foundation for set theory. It is desirable that we do not get carried away by the picture and keep in mind the close connection to mathematical practice on one side, and to general philosophy on the other. (pp. 5-6)[11]

He underlined "a useful heuristic picture" and put in the margin a shorthand note meaning "false" and also underlined "not get carried away by the picture" and put what appears to be a shorthand annotation in the margin.

On the top of page 1 Gödel wrote, "N5, N6 falsch." That stricture might include not only the passages just mentioned but the immediately following long quotation from Marx and the remarks following on it with their emphasis on practice.

This text is not really coherent, and I did not see reason to publish it here. A few years later Wang's interest in Marxism waned along with illusions he had concerning the Chinese regime and conditions there.[12]

§3. *Fragment Q.* I now turn to fragment Q, which is published here. According to Wang's letter to Gödel of December 12, 1975, it was to form the first part of a three-part paper.[13] The section heading "Quotations from Gödel" was intended for the first part of the paper. A number of remarks not quoted from Gödel's writings appear, generally not quite verbatim, as numbered remarks in LJ, but it is very unlikely that the document as we have it consists entirely of literal quotations of utterances of Gödel. However, Wang writes, "Since the whole of (1) is attributed to you, you will undoubtedly wish to examine it thoroughly."[14] Even the question whether the whole of this essay is just an exposition of Gödel's views is not en-

---

[11]It is probably this passage that Wang refers to in the comment on pp. 239-40 of LJ. The following remark 7.4.8 refers to Abraham Robinson as someone holding an "as-if position" and says, as does §1.13 of Q, that the imagination is more successfully stimulated by an objectivist view.

[12]See LJ, p. 123.

[13]Gödel [2003a], p. 414.

[14]*Ibid.*, p. 415.

tirely simple. But for the most part, I will treat the text as if it were written by Gödel.

Sections 2 and 3 were to be "Exposition and commentary" and "Interpretation and broader issues" respectively. So far as is known, they were never written as such. Some of the content Wang envisaged may have been put into [1977a] or other writings, most likely LJ. But it is noteworthy that [1977a] was his last publication specifically devoted to set theory.

The structure of the document is made quite clear by its division into subsections. §§1.1-1.2 concern concepts, first in relation to objects and then in relation to sets. §1.3 introduces a conception of classes as derivative from concepts. §1.4, entitled "Applications," mainly concerns classes and their use in formulating reflection principles; the matter is treated more fully in [1977a]. §1.5 briefly sketches the iterative conception of the universe of sets, relying heavily on quotations from Gödel's published writings. §1.6 returns to the theme of the ontology of sets and their relation to objects more generally. §§1.7-1.8 present arguments for realism. §1.9-1.11 continue some themes on this subject, applying a distinction between an "objective" and a "subjective" point of view. Further arguments, based largely on Gödel's published writings, occur in §§1.12-1.13. §1.14 introduces epistemic considerations and remarks on justifying the axioms of set theory. It introduces the idea of an intuitive range of variability, prominent in chapter VI of FMP; a little more is said on that subject in §1.15, which also remarks on other issues concerning intuition.[15]

§4. *Concept and object, set and class.* Wang begins §1 of [1977a] with the statement

According to Gödel, a set is a unity of which the elements

---

[15]Wang reports (LJ, p. 357) that in December 1975 he sent a copy of fragment Q to Paul Bernays and received a reply with comments. Some of these comments appear as remarks 10.2.6-12 of LJ. As Wang remarks (*ibid.*, p. 358), Bernays evidently does not share Gödel's realism about concepts, and he emphasizes the importance of geometry as a fundamental source for mathematical concepts.
Bernays was an important figure for Wang. On his influence on Wang's approach to the foundations of mathematics, see Parsons [1998], pp. 5-6. Juliet Floyd's essay in this volume brings to light the influence of Bernays on Wang's philosophy more globally.

are constituents. Objects are unities and sets are objects.[16]

In contrast, fragment Q begins with concepts, but the ontological point of view seems to be the same. A concept is said to be a whole composed out of primitive concepts. A set is said to be a unity (or whole) of which the elements are constituents. But for a unity to be a whole, it must be divisible, so that primitive concepts, unit sets, and monads are unities but not wholes.[17] However, in §1.5 he quotes [1944], p. 141, where it is allowed that unit sets and also the null set might be regarded as fictions.[18] At the end of §1.1 it is remarked, "Pure sets are the mathematical objects and make up the world of mathematics." That would imply that other mathematical objects, natural numbers in particular, are reducible to pure sets, and if that were Gödel's view he would probably take it in a metaphysical way. For this reason Wang later doubted that he had got Gödel right on this point; he says that he does not think Gödel wished to deny that numbers are not reducible to sets.[19]

Concepts are said to be more "organic" wholes than sets, but what this amounts to is not explained. However, remarks about sets encourage the conjecture that what he has in mind is that for a set, the only thing that matters is what elements it has, while a concept, although it is composed of certain primitive concepts, is not determined by something like the list of these concepts, but also at least by how they are related.

Gödel seems to think of concepts in a type-free way. This is indicated by the different examples of concepts that he offers, including in particular the concept of concept, and the statement that

---

[16]The first two sentences of fragment M are almost the same, but there instead of simply "unity" Wang writes "unity that is a whole." The omission may have been due to the fact noted in the text, following §1.1 of fragment Q, that unit sets do not count as wholes.

[17]Although monads are mentioned in Gödel's conversations with Wang, apart from this remark they are hardly present in this document.

[18]One might doubt that that was Gödel's considered view. If the null set is a fiction, how can one escape the conclusion that all pure sets are fictions? Also, if x has no elements or more than one, extensionality forces a distinction between x and $\{x\}$, and standard set theory (with foundation), forces the distinction in all cases. It should be noted, however, that in spite of the remark in the text about the role of pure sets, they are less prominent in Gödel's published expositions of the iterative conception than in what is standard today; see especially [1964], pp. 262-63, quoted in §1.5 of Wang's text below.

[19]LJ, p. 296.

86

concepts, in contrast to sets, can apply to themselves. We do not have evidence that Gödel undertook to develop a formal theory of concepts on this basis, but two letters to Gotthard Günther in the 1950s indicate that he was interested in the project.[20]

Apparently missing from Gödel's scheme are propositions. One possibility is that propositions are subsumed under concepts. The idea would be that concepts can combine in various ways, but some combinations have the property that they are true or false without specifying an argument, and these would be propositions. A difficulty with this idea is that Gödel does not mention individual concepts or seem to provide for them. Gödel explicitly affirms that sets are objects but neither affirms nor denies this about concepts. It may be that he was uncertain on the latter point. However, he is reported in remark 7.3.12 of LJ to say that concepts are not objects. But he does say here that they are unities, which might be difficult to reconcile with their not being objects. In §1.3 he says that classes "are pluralities and therefore not objects." However, apparently what they lack is unity, which concepts are said to have. More "structural" issues about sets and concepts will be discussed later.

The statement from Wang [1977a] quoted above and the parallel remark in §1.1 seem to express what I would call an ontological conception of set. In saying that the elements of a set are constituents of it and that a set with more than one element is a whole, he comes close to saying that the elements of a set are parts of it. He no doubt avoids that because he knew that, in the earlier history of set theory, that way of putting things led to confusion between elementhood and inclusion. What is of interest is that Gödel does not show inclination toward a more structuralist view of set theory, which is expressed in my own writings and, somewhat differently, in writings of Geoffrey Hellman, Michael Resnik and Stewart Shapiro, and is implicit in a lot of earlier writing, in particular that of Bernays.[21]

---

[20]See the end of the letter of April 4, 1957 and the letter of January 7, 1959, Gödel [2003], pp. 527-30, 534-35.

[21]See for example Parsons [2008], chs. 3 and 4, Hellman [1989], Resnik [1997], Shapiro [1997], and Bernays [1950]. In [2010], p. 180 n. 30, I wrote that Gödel is "almost entirely silent on issues relating to structuralism." Strictly speaking that is true, since he does not mention structuralist views or address their motivation. But I seem to have forgotten texts like the present one and even what I myself wrote about [1977a] and fragment Q in Gödel [2003a], p. 391.

Another "ontological" dimension of the conception of set expressed here is the idea of sets as "the limiting case of spatiotemporal objects" (§1.5), and "an analogue" of what one would get by thinking of a physical body as completely determined by its parts, quite apart from their relations, or as an analogue of synthesizing aspects to get one object, again leaving out the relations the aspects bear to each other. This does something to elucidate the remark in [1964], p. 272, that the function of the concept of set is synthesis in something like Kant's sense. But Gödel is even more emphatic than in that passage in rejecting Kant's purported subjectivism; it is remarked that "subjective is an euphemism for false."[22]

§1.3 sketches a conception of classes as derivative from concepts. The idea is that talk of classes is only a *façon de parler*, in essence talk of concepts modulo coextensiveness. But to have a theory, the concepts involved are best limited to extensional ones, i.e. concepts C such that if C applies to A, and A and B are coextensive, then C also applies to B.

Gödel appears to be rather hostile to the idea of taking classes as set-like entities apart from concepts. This point of view is expressed in almost the same words in remark 8.6.6 of LJ. He does not consider the option, closer to the actual practice of set theorists, of regarding classes as derivative from predicates, or talk of classes as simply a convenient way of talking about predicates. Even in his published writings, with the exception of the mathematical [1940], much of the work that might be done either by classes or by higher-order logic is done by concepts.

§5. *The paradoxes and the idea of a theory of concepts.* As he had said in print, Gödel does not regard the set-theoretic paradoxes as a serious problem; the iterative conception of set rationalizes set theory in a completely satisfactory way. The emphatic statement of [1964], pp. 262-63, is quoted at the beginning of §1.5. He is also unworried about semantic paradoxes; we defer that for the moment.

---

[22]It is hard to be certain whether Gödel is here interpreting Kant, but the remark seems to express the view that what the fact that, according to Kant, our cognition represents objects as they appear and not as they are in themselves amounts to is that it represents them falsely; cf. Parsons [2010], pp. 175-76. But his way of putting the matter here suggests that he is being more emphatic than he would be in a more considered expression of his view.

Here and in some other places Gödel regards what he calls the intensional paradoxes, concerning the notion of concept and related notions, as unsolved. Much earlier he wrote

> By analyzing the paradoxes to which Cantor's set theory had led, he [Russell] freed them from all mathematical technicalities, thus bringing to light the amazing fact that our logical intuitions (i.e. intuitions concerning such notions as truth, concept, being, class, etc.) are self-contradictory. ([1944], p. 131)

Gödel may still have held in later years the view expressed here; at any rate he was not satisfied with available answers to paradoxes concerning intensional notions, in particular that of concept. The impression one gets from the present text is that these paradoxes are the main obstacle to a satisfactory formal theory of concepts. Solving the paradoxes and constructing such a theory were probably, for him, two faces of the same problem. From the absence of evidence of actual work toward constructing a formal theory of concepts, it is reasonable to conjecture that he did not have an idea for an approach to these paradoxes that satisfied him or even struck him as especially promising.

Gödel is not specific about what paradoxes he has in mind. An intensional version of Russell's paradox would be an obvious one: It seems that one might formulate a concept M that applies to a concept just in case it does not apply to itself, and then M will apply to itself if and only if it does not. In remarks 8.6.24-26 of LJ, Gödel formulates a very similar paradox that has minimal logical presuppositions.

About the semantical paradoxes, he seems by contemporary lights complacent and even naïve. They are disposed of very briefly in §1.2:

> Every definite language must contain only countably many symbols, and no precise language can contain, for example, the word 'true' as applied to all sentences in the language.

Gödel seems to envisage only formalized languages, so that the problem that has occupied some who have thought about semantical paradoxes, to capture the concept of truth that underlies the use of the word 'true' in natural language and our thought about truth when not doing logic, does not move him. The second clause of the above quotation might be taken to allow that a language might contain 'true' applied to its own sentences provided that it allows gaps, an idea that had been applied to the liar paradox in a lot of work in the period leading up to the time of the conversations in connection with which this text was written.[23] But this possibility is not mentioned either here or in related parts of LJ. In general Gödel's view of the liar paradox seems resolutely "Tarskian"; perhaps he was too much impressed with his own discovery, when working on the incompleteness theorem, of what is called Tarski's undefinability theorem. But he also shows a strong bias against the "linguistic turn" in philosophy; one facet of his realism about concepts is the belief that they are quite independent of language. It follows that paradoxes concerning the notion of concept are of a different nature from semantical paradoxes.

§6. *Reflection principles and large cardinals.* Wang [1977a] contains a discussion of strong reflection principles. The set theory presented by Wilhelm Ackermann in [1956] was a starting point. One of its axiom schemata is close to the reflection principle that is provable in ZF, and Ackermann's system proved to be equivalent to ZF. Wang explored the efforts of William C. Powell and W. N. Reinhardt to obtain much stronger theories by introducing reflection-like principles involving higher-order quantification.[24] These would be essentially stronger than the well-known second-order reflection principle of Bernays [1961], which yields inaccessible and Mahlo cardinals and more, but no large cardinals incompatible with $V = L$.

The discussion of [1977a] is in terms of sets and classes, with the possibility admitted (but not developed) that one might have higher-order classes whose elements are classes. In §1.4 of the present document, the point of departure is concepts, in line with Gödel's view expressed in §1.3 that classes are derivative from concepts. Wang in [1977a] probably did not wish to put the exposition in terms of concepts without explicit authorization from Gödel, but

---

[23]Wang probably knew something of the content of Kripke [1975], which was published in December 1975 and had been presented in public lectures in Princeton before. But there is no mention of contemporary work on the semantic paradoxes in LJ and thus no indication that Wang brought it up.

[24]See Powell [1972] and Reinhardt [1974] and [1974a].

he appears also to have had a more mathematical motivation, to understand Reinhardt's proposed axioms without using the intensional ideas that Reinhardt introduced in order to motivate them, in particular the rather speculative idea of "imagined" or "projected" sets.[25]

I described the principles involved as "reflection-like." Wang and Gödel both seemed to understand them as reflection principles, but a significant difference can be seen in the simplest principle formulated by Reinhardt:

$$(\text{R2U}) \qquad (\forall x \in M)(\forall X \subseteq M)[F^{M+}(x, X) \leftrightarrow F(x, jX)].^{26}$$

'M' is ambiguous; it may refer to the class V of all sets, or it may be a set that reflects the universe. In either case one can talk, as (R2U) does, of the class of its subclasses. $M+$, the range of the quantifiers on the left-hand inner formula, is the class of subclasses of M. But j is supposed to be an elementary embedding that maps elements and subclasses of M to a structure containing elements not in M; in fact, restricted to M, j is the identity. Interpreting M as a set, (R2U) implies that its rank is a 1-extendible cardinal, which is the first ordinal moved by j. A 1-extendible cardinal is measurable and has a measure 1 set of measurable cardinals below it.[27] Reinhardt went on to consider stronger extendible cardinals, first by using higher-order versions of his principle.

Reinhardt discussed his ideas with Gödel and testifies that Gödel did not find his justification of extendible cardinals adequate.[28] This is also the opinion of Peter Koellner in his study [2009] of reflection principles. (Because of the postulation of the embedding j, Koellner does not classify Reinhardt's principles as reflection principles.) §7 of that paper gives a lucid discussion of Reinhardt's idea and questions about it.

§7.   *Realism.* The discussion of this topic in §§1.7-1.8 adds little to what we know of Gödel's views from his published writings and those published by the editors of the *Collected Works*. The distinction implicit in other writings of Gödel between arguments

---

[25]I discussed these matters briefly in [1977], pp. 286-88.

[26]This formulation follows Wang [1977a], p. 325. For Reinhardt's formulation see [1974a], p. 196.

[27]See Kanamori [1994], pp. 311-12.

[28]See Reinhardt [1974a], p. 189 n.

for the claim that mathematics has a "real content," which would lead to what William Tait calls realism in the default sense,[29] and arguments for a more strongly realist conclusion, is somewhat less clear here than elsewhere, perhaps because of the brevity of the presentation. But in §1.8, just after presenting considerations of the former kind, he abruptly shifts to a more metaphysical claim, that the fact that "multitudes can be unities" is surprising and seemingly contradictory and is a given, objective fact underlying mathematics. This claim would need a lot of elucidation, which is not given.

The treatment of questions about realism is in my view made less subtle than in Gödel's more careful writings by the reliance on a contrast between his own objectivism and what he calls subjectivism, where that view is little developed and no attempt is made to show that it is a view that anyone has actually held. Elsewhere Gödel uses the terms "subjective" and "subjectivism" in remarks about Kant, but here the term is used with reference to issues remote from Kant, as in the claim at the end of §1.9 that "from a subjective viewpoint" it is hard to find a good reason to correct Frege's "mistaken belief that every concept determines a set."

The "argument from success" of §1.13 is essentially the same as what is presented in the letters to Wang published in edited form in FMP, pp. 8-11. It is not so much an argument for the truth of realism as to the effect that it is a mathematically fruitful view to hold. In the letters Gödel emphasizes his own work, but here he attempts to generalize the idea to that of Paul Cohen and to argue that the latter's work requires that sets be real. I leave it as an exercise for the reader to point out the weaknesses of that argument. However, the parallel remark 8.1.9 of LJ is more nuanced, so the fault may lie more in Wang's reporting.

§8. *Knowledge and intuition.* Epistemic considerations are introduced explicitly at the beginning of §1.14. It is noteworthy that the remarks in these sections generally do not have close parallels in those in LJ. It may be that more of Wang's own thought enters into this part of the text. That is also suggested by the fact that most of §1.14 is taken almost verbatim into Wang [1977a]. It is likely that some remarks are meant to respond to my own criticism in [1977] of Wang's treatment of the axioms of set theory in FMP.

---

[29]See Tait [2005], in particular the Introduction and Essay 4.

Thus he regards speaking of "collecting a multitude of given sets into a set" as dealing with our knowledge about sets, and says the same about "the suggestive term 'genetic concept of set'." My own discussion groups Wang's treatment in FMP with that of some other writers who use these terms without really explaining how literally they are to be taken. Wang certainly did not in that discussion clearly indicate that language of the above sort was meant only to refer to our knowledge and not to anything like the coming into being of sets. Prompted by Gödel, Wang probably thought that making such a distinction would offer a reply to my own remarks about the justification of specific axioms. But the idea is not applied to specific cases either here or in [1977a].

Remarks early in the next section do something to clarify Gödel's terminology concerning intuition. He notes that the German word *anschaulich* is narrower than "intuitive," closer to sense perception, and that it is the narrower notion that is at issue in [1958].[30] However, intuition is "probably narrower than having *Evidenz*," i.e. a proposition's being evident to someone. I do not know what he has in mind by the qualification "probably."

Gödel notes provocatively that intuition is "the opposite of proof" but remarks at the very end of the text that an intuition may be "replaced and substantiated by proofs which may reduce it to less idealized intuitions." Although he has stood out as a friend of mathematical intuition, he has no wish to abandon the mathematician's ideal of proof.

§9. *Intuitive ranges of variability.* On pp. 189-90 of FMP, in a passage clearly attributed to Gödel, five principles for justifying axioms of set theory are listed. The first is

> Existence of sets representing intuitive ranges of variability,
> i.e. multitudes which, in some sense, can be "overviewed."

Wang applies that principle to the cases of the axioms of power set, replacement, and separation in his discussion in FMP, and that is a central object of my criticism. Here he corrects a statement of the earlier discussion (p. 182) that an intuitive range of variability is a necessary as well as sufficient condition for a multitude to be a set. I don't think my criticism depended on the claim Wang here

---

[30]Cf. my [1977], p. 278, and [1995], p. 57-58.

retracts.[31]

At the beginning of §1.15 it is remarked that "to see that the range of a multitude is intuitive is a special case of appealing to mathematical intuition." This is a suggestive but somewhat puzzling remark, since when Gödel speaks of mathematical intuition he generally has in mind something propositional, intuition *that* in the sense I have used elsewhere. There is intuition *of* in Gödel's scheme of things, but the term he usually uses is "perception," in particular of concepts.

Probably echoing Gödel, Wang writes

> The distinction between intuition of objects and intuition of propositions does not seem important. There is no clear separation because in either case something general comes in anyhow. In particular, to see that the range of a multitude is intuitive involves an intuition of both the objects of the multitude and a proposition, viz., that the objects of the multitude form a unity. ([1977a], n. 4, on p. 328)

Taken at face value, this might commit Gödel to the view that in order to know by intuition, let's say, that there is an infinite set, one has to have an intuition or perception of all the elements of the set. As several writers have observed, Gödel in his published remarks about mathematical intuition does not speak of perception of sets, rather of perception of concepts. One might then object that the whole idea of an intuitive range of variability, at least as Wang deploys it, tends to undermine this feature of Gödel's view of intuition.

In Wang's discussion of the axioms and in some of Gödel's related remarks in their conversations, the very strong idealization of the powers of the mind that enters in is driven at least to a considerable extent by the idea Wang expresses in the above quotation, that propositional intuition is inseparably bound up with intuition of objects. Since I have already remarked elsewhere on Wang's reformulations in [1977a], I will not pursue the matter further.[32]

The question is of interest how much the discussion of the axioms of set theory in FMP owes to Gödel. To answer it, it would

---

[31]However, I did quote the "only if" formulation in [1977], p. 275. My criticisms are in that paper, pp. 275-79.

[32]See §2 of my [1998], esp. pp. 10-11.

94

be helpful to have access to a draft of Wang's book as it was at the beginning of his first series of conversations with Gödel in 1971. Wang sent such a draft to Gödel, as he acknowledged in a letter of August 4, 1971.[33] In FMP itself, as we have noted, the principle about intuitive ranges of variability is the first of several attributed to Gödel. (It does not follow that it originated with him; other principles on the list, in particular reflection principles, had been discussed some time before.) Wang wrote later that Gödel found his attempt congenial and "although I do not remember clearly, probably suggested as a way to characterize my approach, the term 'intuitive range of variability'."[34] That would suggest that Wang had the general idea in his earlier draft but did not formulate it as sharply as he could do after Gödel proposed this terminology. That makes it likely that the details concerning particular axioms are entirely due to Wang.

Sets and concepts, on the basis of discussions with Gödel

1. Quotations from Gödel

1.1 *Concepts and objects.* A concept is a whole (a conceptual whole) composed out of primitive concepts such as negation, existence, conjunction, universality, object, the concept of concept, whole, meaning, etc. We have no clear idea of the totality of all concepts. All objects are unities and sets are objects.[35] Every whole is a unity and every unity that is divisible is a whole. For example, the primitive concepts, the null set, the unit sets, and the monads are unities but not wholes. Generally we shall not emphasize the distinction between a unity and a whole. Every unity is something and not nothing; any unity is an entity so that objects and concepts all are entities. A set is a unity (or whole) of which the elements are constituents. A concept is a whole in a stronger sense than a set, it is a more organic whole as a human body is an organic whole of its parts. Objects are in space or close to space. Sets are the limiting case of spatio-temporal objects (see below). Among objects are physical objects and mathematical objects. Pure sets are those sets

---

[33]See Gödel [2003a], p. 410.
[34]LJ, pp. 219-20.
[35]Cf. remarks 8.6.17 and 9.1.26 of LJ.

not involving nonsets so that the only Urelement in the universe of pure sets is the null set. Pure sets are the mathematical objects and make up the world of mathematics.

1.2. *Sets and concepts.* Sets and concepts are so different; their connections are only outward. In particular, no set can belong to itself but some concepts can apply to themselves, for example, the concept of concept (is a concept), the concept of being a concept with no finite extension, the concept of being distinct from ω, the concept of being applicable to more than one entity.[36] Sets are extensions and concepts are [2] intensions. Frege erroneously thought that to each concept there corresponds a set, but there are concepts which correspond to no sets. In fact, none of the examples just given correspond to any set.

Logic studies concepts (intensions). Mathematicians are primarily interested in extensions and we have a systematic study of extensions in set theory which remains a mathematical subject except in its foundations. Mathematicians form and use concepts but they do not investigate generally how concepts are formed. We do not have an equally well developed theory of concepts comparable to set theory and at least at the present stage one does not see how to develop such a theory.

Concerning the paradoxes, the unsolved difficulties are mainly in connection with the intensional paradoxes (such as the concept of not applying to itself) rather than with either the extensional or the semantic paradoxes (see FMP, p. 188 and p. 221). The extensional paradoxes are resolved by a correct understanding of the concept of set, namely the iterative concept of set (see below). With regard to the semantic or linguistic paradoxes, the situation is clear. Every definite language must contain only countably many symbols, and no precise language can contain, for example, the word 'true' as applied to all sentences in the language. Language plays no part in the intensional paradoxes, insofar as they are concerned with concepts which are "properties and relations of things existing independently of our definitions and constructions" ([1944], p. 137).

1.3. *Sets and classes.* Generally the range of applicability of a concept need not form a set, the concept of set being an obvious example. When the range of a concept is a set, the set is its

---

[36]Cf. remark 8.6.3 of LJ.

96

extension. Since [3] strictly speaking an extension should be one
object, we can generally speak of the extension of a concept only as
a *façon de parler*. Bearing this in mind, we can think of classes as
'extensions' of concepts. In a theory of concepts we can introduce
classes by contextual definitions (definitions in use) construed in
an objective sense. They are nothing in themselves and we do not
understand what are merely introduced by contextual definitions
which only tell us how to deal with what are thus introduced according to certain rules. If we are interested in developing a theory
of classes, it is more elegant to deal just with extensional concepts
("mathematical" concepts), viz. the totality T of all concepts such
that if A and B are in T and A applies to B, then A applies to
all concepts in T with the same extension as B. We can then think
of these concepts as classes (so that, for example, the axiom of
'extensionality' holds for classes by design).

Classes appear so much like sets that we tend to forget this
line of thought, which leads from concepts to classes. If we leave
out such considerations, the talk about classes becomes a matter of
make-believe, arbitrarily treating classes as if they were sets again.
Generally classes are pluralities and therefore not objects; only
when they are sets are they unities and objects. Since concepts
can sometimes apply to themselves, their 'extensions' can belong
to themselves; i.e. a class can belong to itself.[37]

It is not evident that every set is the extension of some concept,
but such a conclusion may be provable once we have a developed
theory of concepts and more crucially a more completely developed
theory of sets.[38] We shall assume the conjecture (*): every set is
an extension of some [4] concept. This implies that every set is a
class according to the manner of speaking about classes as though
they were unities. We arrive then at a simplification which enables
us to return to the uncritical way of dealing with classes. And, as
usual, classes which are not sets (e.g. the universe of all sets) can
be referred to as proper classes.

1.4. *Applications (sketch only).* Even though we do not have
a developed theory of concepts, we know enough about concepts
so that we can also have something like a hierarchy of concepts

---

[37]Most of this paragraph and some of the preceding paragraph appear in
remark 8.6.6 of LJ.

[38]These two sentences occur almost verbatim in remark 8.6.14 of LJ.

(and classes) which resembles the hierarchy of sets and contains it as a segment, but does not exhaust all the concepts (or classes). Such hierarchies are peripheral to the theory of concepts since we cannot reach the universe of all classes, which belongs to itself. This promises possibilities of piling hierarchies on hierarchies in the manner suggested by Reinhardt in [1974a]. The difference is [that] we are now on more solid ground and no longer have to speak in terms which are shifting or self-contradictory.

In this way we are, in particular, led to a (partial) theory of classes which exhibits some of our idea of classes and sets as described above. We can take pure sets as the only objects. In order to formalize this theory to the limited extent to which theories of sets and numbers are commonly formalized, it is natural to use the first-order logic with the universe of discourse ranging over classes and the membership relation $\in$ as the single extra relation symbol. Moreover, in order to separate out the sets easily, it is also natural to introduce a constant V standing for the undefined class of all sets. For this theory, any truth Φ about sets expressible in systems like ZF remains true and is expressed by a statement Φ^V which is just the statement Φ relativized to V. One [5] problem is that we do not know much about the proper classes. To consider what other axioms are true for classes, it is more appropriate and more persuasive to think in terms of concepts rather than classes.

From this viewpoint, logic contains mathematics as a proper part under the conjecture (*), and there has been over a long time a confusion between logic and mathematics. Once we use the sharp distinction between sets and concepts, we have made several advances. We have a reasonably convincing foundation for ordinary mathematics according to the iterative concept of set. Going beyond sets becomes an understandable and, in fact, a necessary step for a comprehensive conception of logic. We come back to the program of developing a grand logic except that we are no longer troubled by consequences of the confusion between sets and concepts. For example, we are no longer frustrated by wanting to say contradictory things about classes, but we can now say both that no set can belong to itself and that a concept may apply to itself. In this way, we acquire not only a fairly rich and understandable set theory but also a clearer guidance to our search for axioms which deal with concepts generally. We can examine whether familiar

98

axioms for sets have counterparts for concepts and also examine whether earlier attempts (e.g. in terms of the lambda-calculus and of stratification, etc.) dealing with sets and concepts indiscriminately suggest axioms which are true of concepts generally. Of course, we should also look for new candidates for axioms dealing with concepts. At the present stage, the program of finding axioms for concepts seems to be wide open.

Even without introducing special axioms for proper classes, we can give a satisfactory account of Ackermann's set theory presented in [1956]. [6] The chief new axiom known as Ackermann's principle is justified by the reflection principle that the universe of all sets is undefinable (in the partial theory sketched above without of course using the constant V itself). Originally the conceptual difficulty was a failure to make sense of the proper classes.

Moreover, the (equivalent) theories of Powell (in [1972]) and Reinhardt (in [1974]) can, as they have observed, be formulated more or less equivalently by substituting Ackermann's principle for one of their two crucial new axioms.[39] Let T be the system suggested by Reinhardt as an alternative formulation of his system S* (see [1974], p. 23). Then any natural model of T yields one of S*, and T is equiconsistent with S*. Ackermann's principle in T can again be justified by the reflection principle. The remaining crucial new axiom is then (S3.3) which says roughly that all subclasses of V can be obtained without reference to V. A plausible line for justifying this axiom is to reflect that each set x is small relative to V and that whether x belongs to a subclass of V is a "local" question which can be settled by considering a sufficiently large segment of V. This suggestion as it stands is of course not a proof of (S3.3) but at least points to one type of consideration appropriate for justifying strong axioms of set theory. A similar reformulation of Powell's system is available by substituting Ackermann's principle and an ordinary axiom of extensionality for his extremely strong axiom of extensionality. It should be noted that in these systems we can prove the existence of a lot of measurable cardinals.

---

[39] It is clear that Wang means that Powell's and Reinhardt's theories are equivalent to each other, not that they are equivalent to Ackermann's. As noted in the introduction and in Wang [1977a], Ackermann's system is equivalent to ZF while Powell's and Reinhardt's theories yield large cardinals that are stronger than measurable.

[7] 1.5. *The iterative concept of set.*

> This concept of set, however, according to which a set is something obtainable from the integers (or some other well-defined objects) by iterated application of the operation "set of," not something obtained by dividing the totality of all existing things into two categories, has never led to any antinomy whatsoever; that is, the perfectly "naive" and uncritical working with this concept of set has so far proved completely self-consistent. It follows at once from this explanation of the term "set" that a set of all sets or other sets of a similar extension cannot exist since every set obtained in this way immediately gives rise to further applications of the operation "set of" and, therefore, to the existence of larger sets. ([1964], pp. 262-263)[40]

I "consider mathematical objects to exist independently of our constructions and of our having an intuition of them individually" and "require only that the general mathematical concepts must be sufficiently clear for us to be able to recognize their soundness and the truth of the axioms concerning them" (*ibid.*, p. 262).

One familiar objection to this view is that the null set and unit sets are not adequately accounted for. For example,

> Russell adduces two reasons against the extensional view of sets, namely the existence of (1) the null set, which cannot very well be a collection, and (2) the unit sets, which would have to be identical with their single elements. But it seems to me that these arguments could, if anything, at most prove that the null set and the unit sets (as distinct from their only element) are fictions (introduced to simplify the calculus like the points at infinity in geometry), not that all sets are fictions. ([1944], p. 141)

[8] 1.6. *Sets as objects.* Sets are the limiting case of spatio-temporal objects, either as an analogue of construing a whole physical body as determined entirely by its parts (so that interconnections of the parts play no role) or as an analogue of synthesizing aspects to get one object (with the difference that interrelations of the aspects are disregarded).[41]

---

[40]The final sentence of this quotation is from a footnote attached to the preceding sentence.

[41]Cf. remark 8.2.4 of LJ.

> Evidently the "given" underlying mathematics is closely related to the abstract elements contained in our empirical ideas. Note that there is a close relationship between the concept of set thus explained and the categories of pure understanding in Kant's sense. Namely, the function of both is "synthesis," i.e. the generating of unities out of manifolds (e.g., in Kant, of the idea of *one* object out of its various aspects). It by no means follows, however, that the data of this second kind, because they cannot be associated with actions of certain things on our sense organs, are something purely subjective, as Kant asserted. Rather they, too, may represent an aspect of objective reality, but, as opposed to the sensations, their presence in us may be due to another kind of relationship between ourselves and reality. ([1964], p. 272)[42]

In this context, "objective reality" or "reality" means the external world, viz. the world in space and time. Subjective is an euphemism for false. Thus, Kantian synthesis reveals the independent existence of physical objects rather than merely a subjective creation. In the generation of the idea of *one* object out of its various aspects, if we abstract from the interrelations of the aspects, the one object generated would be the set of which the aspects are constituents provided we thought of these aspects as objects. It is clear that we can likewise synthesize a small number of physical objects into one object (a set) by disregarding all interrelations of these objects. As large [9] (though finite) and infinite magnitudes are envisaged, idealizations (or constitutions of mathematical objects) which go beyond the immediately given are necessary. Husserl speaks of constituting mathematical objects but what is contained in his published work on this matter is merely programmatic. Phenomenological investigation of the constitution of mathematical objects is of fundamental importance for the foundations of mathematics.[43]

1.7. *Argument for objectivism from reference.* There are a number of arguments in favor of objectivism of sets which will be presented here.

---

[42]The sentence "Note that ... aspects." is in Gödel's published text a footnote.

[43]Cf. remark 8.2.8 of LJ. Document no. 013167 ends at this point. No. 013169 begins with the text from p. 9 of that document and then continues, continuing the page numbering as well. What follows is based on that document.

It is only through our knowledge obtained in studying mathematics (and in particular set theory) that the picture of sets existing independently of our knowledge is reached. We know many general propositions about natural numbers to be true (e.g. that there are infinitely many prime numbers) and, for example, we believe that Goldbach's conjecture makes sense, must be either true of false, without room for arbitrariness or conventions. Hence, there must be objective facts about natural numbers. Similarly, we believe the axioms of ZF to be true. Hence, there must be objective facts about sets. But these objective facts must refer to objects which are different from physical objects because, among other things, they are unchangeable in time.

Most of us are inclined to accept the steps in the argument and therefore also the conclusion which, understood together with this argument, does not appear to be excessively strong. Keeping in mind the argument serves to prevent the temptation to misuse the conclusion in ways which go beyond its limited content and open avenues to drawing unwarranted consequences catering to wish fulfillment and worse things. [10]

1.8. *Real content and its irreducibility.* Logic and mathematics (just as physics) are built up on axioms with a real content which cannot be "explained away" ([1944], p. 142). Hence, the so-called "Occam's razor" cannot be applied without contradicting specific evidence referring to sets.

The presence of this real content is seen from studying number theory. We encounter facts which are independent of arbitrary conventions. These facts of mathematics must have a content because the consistency of number theory certainly cannot be based on trivial facts, since it is not even known in the strongest sense of "knowing."[44]

It is a surprising and seemingly contradictory fact that multitudes can be unities (sets). This is the main objective fact which we have not made for mathematics, and it is only the evolution of mathematics which leads us to the discovery of this important

---

[44]This paragraph was changed by crossings-out and handwritten insertions and substitutions from an argument from set theory into an argument from number theory. For details, see the textual notes. Evidently Wang originally intended an argument based on set theory but, very likely prodded by Gödel, replaced it by an argument from number theory.

fact. In the general matter of universals and particulars, we do not have the merge of two opposite things (many being also one) to the extent that multitudes are themselves unities.[45] This significant property of certain multitudes that they are unities must come from some more solid foundation than the apparently trivial and arbitrary phenomenon that we can think together the objects in each of these multitudes. Without the objective picture, we do not seem able to exclude complete arbitrariness in thinking when a multitude can be thought together and when not.

Positivists behave as though they wish to cut part of the brain out: eliminating the general from the actual knowledge we possess and returning to a more primitive stage. But even if we adopt positivism, it seems to me that the assumption of mathematical objects is quite as legitimate as the assumption of physical objects and there is quite [11] as much reason to believe in their existence. They are in the same sense necessary to obtain a satisfactory system of mathematics as physical objects are necessary for physical theory and in both cases it is impossible to interpret the propositions one wants to assert about these objects as propositions about the "data," i.e., in the former case simple numerical calculations and in the latter case the actually occurring sense perceptions. (Compare [1944], p. 137.[46])

1.9. *Unifiability of small multitudes.* J. von Neumann has shown that a multitude is a set if and only if it is smaller than the universe of all sets (i.e. there is no one-one correlation between all sets and objects in the multitude). This is understandable from the objective viewpoint since one object in the whole universe must be small compared to the universe and small multitudes of objects should form unities since being small is an intrinsic property of these multitudes. From the subjective viewpoint, there is little connection between the size of a multitude and thinking together the objects of the multitude in one thought, since a large but homogeneous multitude may hang together in our thought more easily than a small but heterogeneous multitude. For example, from the subjective viewpoint, it is hard to find a good reason to correct Frege's

---

[45]Up to this point, this paragraph is close to remarks 8.2.2-3 of LJ.

[46]The text after "even if we adopt positivism" is close to a direct quotation from the text cited but not an exact quotation.

mistaken belief that every concept determines a set.[47]

1.10. *Objective unity as contained in subjective unity.* If we overview a multitude of objects in one thought in our mind, then this whole contains also as a part the objective unity of the multitude of objects, as well as its relation to our thought. Different persons can each overview the same multitude in one thought. Hence, it is natural to assume a common nucleus which is the objective unity. It is indeed a unity since it is contained in another unity. [12]

1.11. *An argument against an argument from the paradoxes.* The set-theoretical paradoxes are hardly any more troublesome for the objective view of concepts than deceptions of the senses are for physics ([1964], p. 271; compare also FMP, pp. 84-86). The iterative concept of set resolves these extensional (or set-theoretical) paradoxes. With regard to the unresolved intensional paradoxes, the comparison with deceptions of the senses is an adequate argument against the weak argument for the strong conclusion that since there are these (unresolved intensional) paradoxes, concepts cannot exist because existing things cannot be inconsistent in their properties. As noted before (under 1.8), the paradoxes can only show the inadequacy of our perception rather than throw doubt on the reality of the subject matter. On the contrary, they reveal something which is not arbitrary and can, therefore, also serve to suggest that we are indeed dealing with something real. Subjective means that we can form concepts arbitrarily by subjectively correct principles of formations of thought. Since the principles leading to the paradoxes seem to be quite correct in this sense, the paradoxes prove that subjectivism is mistaken.[48]

1.12. *Objects and objectivity.* There are intuitions and potential arguments which lead us to believe that certain mathematical axioms are true or that certain undecided propositons are meaningful (i.e. true or false). These do not directly argue for the existence of mathematical objects. But by the argument from reference (1.7 above), truth or meaningfulness implies the existence of objective facts which must refer to objects. We give here one such argument and one such potential argument.

---

[47]This section is tracked closely by remark 8.3.7 of LJ.

[48]From "With regard to the unresolved intensional paradoxes" to the end of the section is tracked pretty closely by remark 7.4.5 of LJ.

But even if we don't believe in the "truth" of set-theoretical [13] intuition,

> The mere psychological fact of the existence of an intuition which is sufficiently clear to produce the axioms of set theory and an open series of extensions of them suffices to give meaning to the question of the truth or falsity of propositions like Cantor's continuum hypothesis. What, however, perhaps more than anything else, justifies the acceptance of this criterion of truth in set theory is the fact that continued appeals to mathematical intuition are necessary not only for obtaining unambiguous answers to questions of transfinite set theory, but also for the solution of the problems of finitary number theory (of the type of Goldbach's conjecture), where the meaningfulness and unambiguity of the concepts entering into them [and the higher and higher axioms leading to their solution] can hardly be doubted. This follows from the fact that for every axiomatic system there are infinitely many undecidable propositions of this type. ([1964], p. 272)

The potential argument for the truth of the axioms of set theory depends on their possible elementary consequences.

> Besides mathematical intuition, there exists another (though only probable) criterion of the truth of mathematical axioms, namely their fruitfulness in mathematics and, one may add, possibly also in physics. This criterion, however, though it may become decisive in the future, cannot yet be applied to the specifically set-theoretical axioms (such as those referring to great cardinal numbers), because very little is known about their consequences in other fields. The simplest case of an application of the criterion under discussion arises when some set-theoretical axiom has number-theoretical consequences verifiable by computation up to any given integer. On the basis of what is known today, however, it is not possible to make the truth of any set-theoretical axiom reasonably probable in this manner. (*ibid.*)

Note that this criterion is the same as that which, under general agreement, the truth of physical theories is decided. [14]

1.13. *Argument from success.* My objectivistic conception of mathematics and metamathematics in general, and of transfinite reasoning in particular, was fundamental to my work in logic (FMP, p. 9; for detailed explanation, see *ibid.*, pp. 8-11). My completeness theorem of the predicate calculus [1930] is, mathematically, an almost trivial consequence of Skolem's work in 1922.[49] The explanation of this strange failure of Skolem and others to get the theorem lies in a widespread lack, at that time, of the required epistemological attitude toward metamathematics and toward nonfinitary reasoning. The heuristic principle of my construction of undecidable number theoretical propositions in the formal systems of mathematics is the highly transfinite concept of "objective mathematical truth" as opposed to that of "demonstrability." The ramified hierarchy, which had been invented expressly for constructivistic purposes, has to be used in an entirely nonconstructive way in my consistency proof for the continuum hypothesis.

The work of Paul J. Cohen is based on my development of constructible sets. (In fact, I had previously in unpublished work developed part of Cohen's method and proved the independence of the axiom of choice.) In Cohen's independence proofs, one makes statements about what one does not know. This would be nonsense if sets were not real but were only as one constructs them oneself. Only because sets are real, you can make definite statements about them even though you only know them to a small extent. The following fact is clear in these proofs. If an arbitrary set is envisaged, empirical knowledge cannot dispose of it by defining a definite limit; but more statements can be made about it.[50]

Nobody denies the fact that in set theory one at least speaks as if there are sets and that psychologically such an objectivist picture is [15] of help in studying set theory. It is sometimes suggested that one only has to pretend that sets exist. But by such pretending one

---

[49]I.e. Skolem [1923].

[50]Cf. remark 8.1.9 of LJ. However, this whole paragraph has a bracket in the left margin with a shorthand symbol meaning *falsch*. This may refer to the parenthesis, where Gödel is reported to make a more definite claim about his accomplishments on independence questions than he was willing to make in other contexts. See the 1966 portion of the correspondence with Alonzo Church in Gödel [2003] and the correspondence with Wolfgang Rautenberg in [2003a], together with the introductory notes. Cf. also the statement in Wang [1981c], p. 657, which also claims more for Gödel.

can never arrive at the same degree of imagination as is obtained by some objectivists. Success in the applications is the usual and most effective way of proving existence. It would indeed be strange that one gets success from a wrong picture which is a mere pretension rather than from the real situation.

The notion of existence in the most general sense (in the weakest sense) is one of the primitive concepts with which we must begin and which must be taken as given. It is the most clear abstract concept we have. Even the concept of "all" is not as clear. Hence the tendency of positivists to ask the meaning of existence (in particular, of mathematical objects) is directed at a wrong question.

1.14. *Knowledge and existence.* Thus far we have tried to keep the knowing subject out as far as possible and to reach the iterative concept of set without reference to any temporal or subjective priority. Once we have arrived at this picture of what is commonly known as the rank hierarchy, we can conveniently speak of collecting a multitude of given sets into a new set. This now deals with our knowledge about sets and does not concern the ontological status of sets. Also, the suggestive term "genetic concept of set" belongs to the same level (viz. of knowledge rather than of existence).

Of course, for our knowledge, there are many difficulties about this iterative concept or rank hierarchy. We are led to the objective picture by extrapolation from our limited knowledge, and once we get the objective picture which is necessarily blurred in places, we are obliged [16] to be more explicit in order to convince ourselves that axioms of set theory are true for this picture. For example, it is left vague as to how far we can continue the iteration process, assuming implicitly that we continue "as far as possible." This brings us to the question of ordinal numbers as the indices of the stages of iterations. Of course, we do not assume all ordinals given in advance since, for our knowledge, they do not come from heaven. Rather the formation of sets and stages can interact so that, for example, the axiom of replacement can yield sets which lead to new stages.

Generally it should be clear that there are diverse ways of introducing and justifying axioms. The justifications can only to a greater or lesser degree serve to indicate that the axioms under consideration are true for the iterative concept of set. Elsewhere I have listed a heterogeneous and not mutually exclusive group of five ways of justifying axioms of set theory. (Wang [1974a], pp. 189-190.) Of these five ways, the first one (viz. that of overviewing a multitude as an intuitive range of variability) occupies a special place and is the only one acceptable to the subjectivists. It is sufficient to yield enough of set theory as a foundation for classical mathematics and has in fact been applied (in *ibid.*, pp. 181-190) to justify all the axioms of ZF, which go beyond what is needed for classical mathematics. In contrast with it, the other four ways (including, in particular, the reflection principle) are more objective and may be spoken of as being abstract rather than intuitive. For example, the outline in 1.6 above of justifying Ackermann's principle by the reflection principle goes beyond the use of intuitive ranges.

The word "only" should be deleted from the statement "we can form [17] a set from a multitude only in case the range of variability of this multitude is in some sense intuitive" (*ibid.*, p. 182). Otherwise it would be too strong and imply that the other four ways of justifying axioms of set theory all are reducible to the way of discovering that the ranges of variability are intuitive. In applying the idea of having an overview of a range, we are helped by a contrast between the power set of ω and the totality V of all sets obtainable by the iterative concept. Admittedly the intuition of the former calls for a strong idealization as is seen from the fact that we have no constructive consistency proof of the assumption of its existence. But clearly we do not have even a similarly weak intuition of the range of V.[51] The uneasiness toward the power set of ω is often a result of having in mind definitions rather than objects. The transcendental approach requires of those who recognize the existence of set theory as a fact to assent to the conditions of its possibility which are not satisfactorily accountable from a positivistic position. I wrote certain passages of my philosophical article [1964] on Cantor's problem for the[52] practical purpose of driving away from mathematicians the fear of doing set theory because of the

---

[51]This section up to this point is tracked almost verbatim by pp. 312-13 of Wang [1977a].

[52]In the right margin is a shorthand annotation, transcribed by Cheryl Dawson as *Was heißt das?* Gödel seems to be asking what was the basis for the assertion about his earlier intentions. However, a similar statement, now referring to the entire paper, occurs as remark 8.1.12 of LJ.

Although Wang cites [1964], Gödel would presumably have had in mind already [1947].

paradoxes.

1.15. *Intuitive range and intuition.* To see that the range of a multitude is intuitive is a special case of appealing to mathematical intuition. Intuition is the opposite of proof, it is to see something without a proof.[53] We undertake to describe what we see which generally cannot be analyzed further (in particular, so as to see a proof). The description is intended to invoke the intuition. The German word *anschaulich* is more closely related to sense perception and concrete objects and has a narrower range of application than "intuitive" which is probably narrower [18] than having *Evidenz*. (For example, the consideration of the finitist viewpoint in [1958] is concerned with *Anschauung*.) Intuition has been idealized to different extents. G. A. Miller observes that we can grasp all at once only about seven items plus or minus one or two (see his [1956]). Bernays notices that from two integers k and m one passes immediately to k^m and that this process leads in a few steps to numbers which are far greater than any occurring in experience ([1935], trans. p. 280). Brouwer idealizes our mathematical intuition at least to the extent of asserting that recursive operations on arbitrarily large integers are intuitive. Poincaré and Weyl go beyond Brouwer in permitting nonrecursive but predicative formation of sets (in particular, those of integers). The objectivist position goes far beyond in idealizing intuition. Physics has eliminated its former dependence on some of the more general intuitions such as the acceptance on intuitive grounds that space is Euclidean. It is possible that as mathematics progresses, we shall be able to eliminate some of the extensive idealizations of our mathematical intuition to which we appeal at present.

The arguments for mathematical objectivism given above are admittedly not decisive and insofar as they are decisive do not prove either Platonism in its original sense or any of the external applications and misapplications traditionally associated with Platonism. To apply a position beyond its limits of validity is the most vicious way of discrediting it. This is also true of the emphasis on intuition: appealing to intuition calls for more caution and more experience than the use of proofs, not less. While appeal to intuition continues to be necessary, it is always a step forward when an intuition

---

[53]Cf. remark 9.2.46 of LJ.

(or part of it) is replaced and substantiated by proofs which may reduce it to less idealized intuitions.[54]

---

[54]From "To apply" on this paragraph agrees almost verbatim with remark 9.2.24 of LJ.

Textual Notes

(Places in the text are identified by subsection number, paragraph, and line. Readings of Gödel's shorthand are due to Cheryl Dawson, except in specific places where it is due to Robin Rollinger.)

At the top left of the first page is written 'p. 4' followed by shorthand reading "oben Fortsetzung." At top middle is written 'iterat conc of set' followed by shorthand. Cheryl Dawson originally read it as "die einzige Weise." She has come to doubt "Weise" and now suggests 'wissen', whether as noun or verb. That leaves uncertain how the whole remark is to be read.

1.1, 1, 10. Comma after 'nothing' changed to semicolon.

1.2, 1, 2. 'outwardly' corrected to 'outward'.

1.2, 1, 4. 'a' in 'is a concept' apparently underlined.

1.2, 3, 5-6. The words 'a correct understanding of the concept of set, namely' are a handwritten insertion.

1.2, 3, 7. Comma added after 'paradoxes'.

1.3, 1, 6. Italics of *façon de parler* are due to the editor.

1.3, 1, 10. Above the word 'what' is a handwritten 'but'. Substituting it would make the sentence ungrammatical, but it would become grammatical, and perhaps yield improvement, by omitting 'and we do not understand'. No such editing is undertaken in the parallel remark 8.6.6 of LJ.

1.3, 2, 2. Comma inserted after 'thought'.

1.5, 1, 8 of quote. self-consistent. 'self-', missing from Wang's quotation, has been restored from Gödel's text.

1.5, 2, 1 and 3. 'I' is handwritten and replaces a crossed-out typed 'Gödel'. The verbs 'consider' and 'require' have crossed-out 's's which would have marked them as third person singular, as they are in Gödel's published text.

1.6. Because the quotation is not set off in the original typescript, the division into paragraphs is due to the editor.

1.6, 1, 1-3. 'Sets are ... parts' is underlined by hand. In the right margin is shorthand reading "falsch."

1.6, 2, 3. 'ideas' corrects the typescript, which inaccurately gives 'idea'.

1.6, 3, 15-16. 'investigations' corrected to 'investigation' to make it agree with the verb 'is'. This grammatical error occurs in the parallel remark 8.2.8 of LJ.

First page of no. 013169 (top of p. 9 of typescript): On left 'p. 17 I. Person' (first person).[55] To right 'p. 9, p. 10', then 'p. 14', in separated off part with shorthand, of which Dawson suggests the reading, "Er hätte sollen wörtlich doch alles wie in der ursprünglichen Version nur einige kleine aus Bedrückten und add." ("He should have [included] literally everything in the original version, only a few small [things] from printed [things] and additions.") The word 'version' is in longhand.[56]

Below this text on left:

> intersperse iness[ential]? things & distract the att[ention]
> on the other hand the main points of arg[ument] often concealed.

(What is rendered 'on the other hand' is shorthand reading "andererseits.")

1.7, 2, 3-5. The sentence beginning 'We know' up to 'for example,' is enclosed in handwritten square brackets.

1.7, 2, 6-7. 'without room for arb[itrariness] or conventions' is a handwritten insertion. 'arb' could be read 'arbitrary', but that would make it difficult to construe the sentence.

1.7, 2, 8. Above 'we' is written 'some', but 'we' is not crossed out.

1.7, 2, 8-10. The two sentences 'Similarly, ... true' and 'Hence ... sets' are enclosed in handwritten square brackets.

1.7, 3, 3. The word 'excessively' is underlined by hand, and above it is a handwritten question mark.

1.7, 3, 5-7. In the left margin next to the last two lines of the typescript page (from 'go beyond' to 'worse things') is a handwritten question mark. The words 'to wish fulfillment' are apparently underlined by hand.

---

[55]This is somewhat mysterious but may refer to the sentence at the end of §1.14, which he queried; see below.

[56]This and the following remark parallel a remark that Gödel made on the envelope in which Wang sent the second installment of our text. See §2 of the introduction.

112

1.8, 1, 1. Above 'Logic and mathematics' is written (apparently in Gödel's hand) 'no argument'.

1.8, 1, 3-5. This sentence is enclosed in handwritten square brackets, and the words 'Occam's razor' and 'evidence' are underlined by hand.

1.8, 2. In the typescript, without handwritten changes other than the obvious correction of 'contact' to 'content', this paragraph reads:

> The presence of this real cont[ent] is seen from studying set theory. We encounter facts which are independent of arbitrary conventions. These facts must have a content because the consistency of set theory is certainly not trivial, since it is not even known in any strong sense of 'knowing'. Moreover, we cannot assume sets arbitrarily because otherwise we get contradictions,

Above 'set theory' is written 'no. theory' (possibly in Gödel's hand); 'facts of math[ematics]' is a handwritten insertion; in the same sentence 'set' is crossed out and 'number' written above; 'cannot be based on trivial facts' is a handwritten correction of 'not trivial'; 'the strongest sense' results from handwritten corrections (possibly in Gödel's hand) of 'any strong sense'. The final typed sentence is crossed out and marked 'different argument'. Below 'otherwise' is shorthand probably reading "kommt etwas anderes vor."

1.8, 3, 1. '& seemingly contradictory' is a handwritten insertion, 'contradictory' apparently in Gödel's hand.

1.8, 3, 2-3. 'objective' and 'which *we* have not made' are handwritten insertions. The resulting sentence is somewhat awkward; it might be better phrased as, "This is the main objective fact for mathematics, one that *we* have not made."

1.8, 3, 10. 'think together' is handwritten, above the crossed-out 'overview'.

1.8, 3, 12-13. 'thinking' is handwritten, replacing crossed-out 'determining'. 'thought together' replaces the crossed-out 'overviewed'. However, 'overview/ed' is retained in the parallel remark 8.3.1 of LJ.

At the end of the paragraph is added 'but if we do we arrive at contradictions'. This doesn't fit into the sentence, but the meaning is apparently that we arrive at contradictions if we fail to "exclude complete arbitrariness."

1.8, 4, 3-4. 'But' is a handwritten insertion. 'positivism' is a handwritten replacement for the crossed-out 'the language of positivists'.

1.10, 1, 2. 'overview' is underlined and 'attend on' is written above it. I have left the typed text because it is not crossed out. In any case 'attend to' would be better English.

1.11, 1, 3. The word 'mathematics' is crossed out and replaced by 'the obj. view of conc.' (I am not sure of the reading of the last word as 'conc.', but I do not see what other reading would make sense.) Without this change, the text would be an exact quotation from the cited place in Gödel [1964].

1.11, 1, 4. Above 'physics' is written 'the obj. view of phy', but 'physics' is not crossed out.

1.11, 1, 4-5. 'Moreover' at the beginning of the sentence is crossed out. At the end of the sentence is a handwritten insertion 'exactly as phy e.g. opt' followed by a letter that appears to be 'h'. I don't know what it could abbreviate.

1.11, 1, 9. 'concepts' apparently replaced 'the concept of concept' in the typescript: 'the' is apparently crossed out; 's' is added by hand, and a following 'of concept' is crossed out.

1.11, 1, 10-11. The ending clause beginning 'because' replaces crossed-out 'so that it is impossible to arrive at a serious theory of concepts'.

1.11, 1, 12. 'our perception' is a handwritten insertion replacing the crossed-out 'existing theories'.

1.11, 1, 16. 'subjectively' is a handwritten insertion.

1.11, 1, 17. 'formation' in the typescript is crossed out; 'of formations of thought' is inserted by hand. Clearly the extra 'of' was a slip.

1.11, 1, 18. 'seem to be quite correct in this sense' replaces the crossed out 'are unavoidable'.

1.12, 1, 1. 'intuitions' replaces the crossed-out 'arguments'. 'intuition' might be a possible reading, but it does not make grammatical sense.

1.12, 1, 4-5. This sentence is enclosed in handwritten square brackets, and 'do not directly' is apparently underlined by hand. A close bracket occurs just after the following 'But'; very likely it was written by mistake.

1.12, 1, 7-8. The sentence 'We give here one such argument ...' may have been crossed out by hand, but the intention may have been just to underline it.

1.12, 2, 1-2. This phrase, apparently in Gödel's hand, replaces the crossed-out 'The argument is based on the fact of the existence of mathematical intuition'. Gödel's word 'intuition' is redundant, because the typed 'intuition' is not crossed out.

1.12, 2, 13-14 of quote. 'and the higher & higher ax. leading to their sol.' is handwritten just after the end of the paragraph, marked to be inserted at the place in the quotation where it occurs above. This appears to be in Gödel's hand, and the insertion is in all probability due to him; the sign for the insertion is like those in the manuscript of the Gibbs lecture.[57] Below 'their sol.' is shorthand reading "Analogie mit Mengenlehre."

1.12, 3, 1. 'the truth of' is a handwritten insertion.

1.12, 4. In the right margin, near the beginning of the quotation, is shorthand probably reading 'auf einem höheren Niveau'.[58]

1.12, 4. Sentence after quote. This sentence is a handwritten addition, apparently in Gödel's hand.

1.13. The subsection number 1.12 is repeated here; this has been corrected from here on.

1.13, 2. This paragraph is marked in the margin with a shorthand annotation meaning "falsch"; see note 44 to the text.

1.14, 1. A shorthand comment reading "hier Fortsetzung" immediately follows this paragraph.[59] A handwritten slash symbol '/' may indicate that something that would be the "continuation" was to be inserted between here and the next paragraph.

1.14, 4, last 4 lines. The sentence from 'I wrote' to 'on Cantor's problem for the practical' is underlined, probably by Gödel. In the right margin is a shorthand annotation reading "Was heisst das?".

1.15, 2, last 4 lines. From 'is replaced' to the end of the sentence underlined by hand, probably by Gödel.

---

[57]See Gödel [1995], pp. 288-89.

[58]This reading is due to Robin Rollinger.

[59]This reading is due to Rollinger.

References

Cited works of Gödel:
(Gödel's publications are cited in the original pagination, given in the margins of the *Collected Works*.)

1930. Die Vollständigkeit der Axiome des logischen Funktionen-kalküls. *Monatshefte für Mathematik und Physik* 37, 349-360. Reprinted with English translation in Gödel [1986].

1940. *The Consistency of the Axiom of Choice and of the Generalized Continuum Hypothesis with the Axioms of Set Theory*. Princeton University Press. Reprinted in Gödel [1990].

1944. Russell's mathematical logic. In Paul Arthur Schilpp (ed.), *The Philosophy of Bertrand Russell*, pp. 125-153. Evanston: Northwestern University. Reprinted in Gödel [1990].

1947. What is Cantor's continuum problem? *American Mathematical Monthly* 54, 515-525. Reprinted in Gödel [1990].

1958. Über eine bisher noch nicht benützte Erweiterung des finiten Standpunktes. *Dialectica* 12, 280-287. Reprinted with English translation in Gödel [1990].

1964. Revised and expanded version of Gödel [1947]. In Benacerraf and Putnam [1964], pp. 258-273. Reprinted in Gödel [1990].

1986. *Collected Works*, volume I. *Publications 1929-1936*. Edited by Solomon Feferman, John W. Dawson, Jr., Stephen C. Kleene, Gregory H. Moore, Robert M Solovay, and Jean van Heijenoort. New York and Oxford: Oxford University Press.

1990. *Collected Works*, volume II. *Publications 1938-1974*. Edited by Solomon Feferman, John W. Dawson, Jr., Stephen C. Kleene, Gregory H. Moore, Robert M. Solovay, and Jean van Heijenoort. New York and Oxford: Oxford University Press.

1995. *Collected Works*, volume III. *Unpublished Essays and Lectures*. Edited by Solomon Feferman, John W. Dawson, Jr., Warren Goldfarb, Charles Parsons, and Robert M. Solovay. New York and Oxford: Oxford University Press.

116

2003. *Collected Works*, volume IV. *Correspondence A-G*. Edited by Solomon Feferman, John W. Dawson, Jr., Warren Goldfarb, Charles Parsons, and Wilfried Sieg. Oxford: Clarendon Press.

2003a. *Collected Works*, volume V. *Correspondence H-Z*. Edited by Solomon Feferman, John W. Dawson, Jr., Warren Goldfarb, Charles Parsons, and Wilfried Sieg. Oxford: Clarendon Press.

Other works cited:

Ackermann, Wilhelm, 1956. Zur Axiomatik der Mengenlehre. *Mathematische Annalen* 131, 336-345.

Benacerraf, Paul, and Hilary Putnam (eds.), 1964. *Philosophy of Mathematics: Selected Readings*. Englewood Cliffs, N.J.: Prentice-Hall. 2nd ed., Cambridge University Press, 1983.

Bernays, Paul, 1935. Sur le platonisme dans les mathématiques. *L'enseignement mathématique* 34, 52-69. English translation in Benacerraf and Putnam [1964].

Bernays, Paul, 1950. Mathematische Existenz und Widerspruchsfreiheit. In *Études de philosophie des sciences, en hommage à F. Gonseth à l'occasion de son soixantième anniversaire*, pp. 11-25. Neuchâtel: Griffon, 1950. Reprinted in *Abhandlungen zur Philosophie der Mathematik* (Darmstadt: Wissenschaftliche Buchgesellschaft, 1976), pp. 92-106.

Bernays, Paul, 1961. Zur Frage der Unendlichkeitsschemata in der axiomatischen Mengenlehre. In Yehoshua Bar-Hillel, E. I. J. Poznanski, M. O. Rabin, and Abraham Robinson (eds.), *Essays on the Foundations of Mathematics, Dedicated to Prof. A. A. Fraenkel on his Seventieth Anniversary*, pp. 3-49. Jerusalem: Magnes Press.

Butts, Robert E., and Jaakko Hintikka (eds.), 1977. *Logic, Foundations of Mathematics, and Computability Theory*. Dordrecht: Reidel.

Hellman, Geoffrey, 1989. *Mathematics without Numbers. Toward a Modal-Structural Interpretation*. Oxford: Clarendon Press.

Kanamori, Akihiro, 1994. *The Higher Infinite. Large Cardinals in Set Theory from their Beginnings*. Springer-Verlag.

Koellner, Peter, 2009. On reflection principles. *Annals of Pure and Applied Logic* 157, 206-219.

Kripke, Saul, 1975. Outline of a theory of truth. *The Journal of Philosophy* 72, 690-716.

Miller, George A., 1956. The magical number seven, plus or minus 2. *Psychological Review* 63, 81-97.

Parsons, Charles, 1977. What is the iterative conception of set? In Butts and Hintikka [1977], pp. 335-367. Reprinted in *Mathematics in Philosophy*. Ithaca, NY: Cornell University Press, 1983. Cited according to reprint.

Parsons, Charles, 1995. Platonism and mathematical intuition in Kurt Gödel's thought. *The Bulletin of Symbolic Logic* 1, 44-74.

Parsons, Charles, 1998. Hao Wang as philosopher and as interpreter of Gödel. *Philosophia Mathematica* (3) 6, 3-24.

Parsons, Charles, 2008. *Mathematical Thought and its Objects.* Cambridge University Press.

Parsons, Charles, 2010. Gödel and philosophical idealism. *Philosophia Mathematica* (3) 18, 166-192.

Powell, William C., 1972. Set theory with predication. Ph.D. dissertation, State University of New York at Buffalo.

Reinhardt, W. N., 1974. Set existence principles of Shoenfield, Ackermann, and Powell. *Fundamenta Mathematicae* 84, 12-41.

Reinhardt, W. N., 1974a. Remarks on reflection principles, large cardinals, and elementary embeddings. In Thomas Jech (ed.), *Axiomatic Set Theory*, part 2, pp. 189-206. Proceedings of Symposia in Pure Mathematics, vol. 13, part 2. Providence: American Mathematical Society.

Resnik, Michael, 1997. *Mathematics as a Science of Patterns.* Oxford: Clarendon Press.

Shapiro, Stewart, 1997. *Philosophy of Mathematics: Structure and Ontology.* New York and Oxford: Oxford University Press.

Skolem, Thoralf, 1923. Einige Bemerkungen zur axiomatischen Begründung der Mengenlehre. In *Mathematikerkongressen i Helsingfors den 4-7 Juli 1922, Den femte skandinaviska mathematikerkon-*

118

gressen, Redogörelse, pp. 1-36. Helsinki: Akademiska Bokhandeln. Reprinted in *Selected Works in Logic*, Jens Erik Fenstad, ed., Oslo: Universitetsforlaget, 1970.

Tait, William, 2005. *The Provenance of Pure Reason. Essays on the Philosophy of Mathematics and its History*. New York and Oxford: Oxford University Press.

Wang, Hao, 1974a. *From Mathematics to Philosophy*. London: Routledge and Kegan Paul.

Wang, Hao, 1977a. Large sets. In Butts and Hintikka [1977], pp. 309-333.

Wang, Hao, 1981c. Some facts about Kurt Gödel. *The Journal of Symbolic Logic* 46, 653-659.

Wang, Hao, 1996a. *A Logical Journey: From Gödel to Philosophy*. Cambridge, Mass.: MIT Press.

Hao Wang on the Cultivation and Methodology of Philosophy

Abner Shimony

1. Introduction

Hao Wang is generally recognized by experts as one of the leading mathematical logicians of the latter part of the twentieth century and one of the deepest philosophers of mathematics, logic, and related subjects. Other commentators are more knowledgeable than I about these central aspects of his work. I shall discuss instead his contributions to general philosophy, which most philosophers seem to have neglected (notable exceptions being Koehler [1998] and Parsons [1998]). The four books—*From Mathematics to Philosophy* ([1974a], cited as FMP), *Beyond Analytic Philosophy* ([1985a], cited as BAP), *Reflections on Kurt Gödel* ([1987a], cited as RKG), and *A Logical Journey: from Gödel to Philosophy* ([1996a], cited as LJ)—are in my opinion a treasure of reflections not only on the subjects of his greatest expertise, but also on philosophy of mind, philosophy of language, epistemology, metaphysics, philosophy of religion, ethics, political philosophy, and philosophical methodology.

For a number of reasons Wang's remarks on metaphilosophy and philosophical methodology are particularly valuable. In the first place, as befits some one who was both a logician and a man of comprehensive interests, Wang was concerned to think systematically about every topic that he treated, even though he made no pretense of achieving his own comprehensive world view. Second, he was a dissident among those twentieth century philosophers for whom logic was the central philosophical discipline, sharply criticizing (especially in *Beyond Analytic Philosophy*) the methods and the substantive positions of most of the analytic and linguistic philosophers of the epoch. Third, Hao Wang was for many years the confidant, in conversations which he characterized as "informal and loosely structured" (LJ, p. 5), of Kurt Gödel, who also dissented from the analytic philosophy of his time. (Gödel's acceptance of Wang as his confidant must have been due not only to his respect for Wang's logical expertise but also to the recognition that Wang would be sympathetic and attentive to Gödel's philosophical ideas.[1]) Gödel spoke in detail about philosophical ideas that he had not published and often did not feel ready for publication. Wang was not passive in these conversations but asked questions, raised objections, and presented ideas of his own. Their conversation was indeed a model of philosophical dialectic. A central and recurrent topic in this dialectic was philosophical methodology, understandably against the background of logic and mathematics.

2. The cultivation of philosophy

Before attempting a summary and analysis of Wang's proposals on philosophical methodology I wish to point out something remarkable and rather rare in the history of philosophy: his remarks on the *cultivation* of philosophy, which concern culture and attitude rather than rules and principles. Of course there were predecessors, such as Marcus Aurelius, Boethius, and Montaigne, who were much concerned with the cultivation of philosophy as a guide to life, but they were moralists and not comprehensive philosophers. What is rare is Wang's reflection on the cultivation of philosophy as a propaedeutic to philosophy as a technical discipline.

There are some personal elements in Wang's discussion of the cultivation of philosophy, of which all are biographically interesting and most are amenable to generalization. He is, first of all, the

---

[1]Charles Parsons, in correspondence, wrote: "I don't know much of any evidence about what Gödel knew in 1967 about Wang's philosophical views. He clearly found interesting the questions Wang raised in a letter of 19 December 1967 (Gödel [2003], pp. 399-403), but I don't know of any reactions to philosophical publications of Wang before he began commenting on drafts of *From Mathematics to Philosophy*. He may well, however, have heard relevant things from Kreisel, Bernays, and possibly others. He had probably at least read Wang [1958a], since it appeared in the Bernays Festschrift to which he himself contributed."

product of two different cultures.

> ... my professional training is nearly all in Western philoso-
> phy (much of it even logic-oriented), yet my formative years
> were lived in China. I have tried hard but have not been
> able to shake off my early conviction that philosophy is not
> just one subject more or less like any other, but something
> special. I continue to believe that philosophy should some-
> how be comprehensive and aim at a unified (and preferably
> with a moderate degree of structure) outlook. In particular,
> it should find a place for each of the different types of em-
> phases I am familiar with, for example, by understanding
> the sources of differences and disagreements. I find myself
> attached to the Chinese tradition of mixing together phi-
> losophy, literature, and history; the interest in politics ties
> it to history, the interpretation of texts merges philosophy
> with its history, and the concern with the unity of nature
> and person overlaps with art and literature. (BAP, p. 194)

A second element, which certainly is in part a consequence of the
first, is his intense social concern, which took the form between 1972
and 1979 (LJ, pp. 123-124) of an interest in Marxism, especially its
Maoist variant, and later of a critical reaction to this attraction.
A third element is Wang's training and professional immersion in
mathematics and logic, which provided standards of rigor, clarity,
depth, fertility, and intellectual beauty impossible to forget in his
studies of other subjects in spite of his realization that they are
not always applicable without appropriate mitigation. A fourth
element, mentioned previously, is Wang's close relation to Kurt
Gödel, whose great work in mathematical logic was followed by
many years of concentration upon philosophy. A fifth element is
something which Wang could show but not say in his writings, but
was obvious to those who knew him personally—the remarkable
mingling of fineness, courtesy, modesty, confidence, strength, and
tenacity in his character.

Out of these personal elements were distilled certain attitudes
toward the cultivation of philosophy. He was acutely aware of philosophical diversity: "Philosophy has a longer history and a larger
diversity of traditions to select from. Related to this is the indef-
initeness of the scope of philosophy and its relevance which leaves
more room for particularizing factors to operate" (BAP, p. 192). He

had a student's attitude towards the history of philosophy, which he explained by saying, "Since ... I do not have as strong convictions as he [Gödel] does on most of the fundamental issues in philosophy, my situation is closer to that of a beginner in philosophy who tries to learn from alternative philosophies by checking what they say against what we suppose we know" (LJ, p. 327). He was certain of the indispensability of clear thinking for philosophy, and though he seems willing to admit that expertise in various disciplines may promote clear thinking, he has no doubt at all that mathematics has this function: "Philosophy sees in mathematics a model of clear thinking, with sharp consequences and indubitable conclusions as well as a universe of discourse in which order prevails and the power of pure reason is most impressive" (LJ, p. 15).

From among the historically enunciated aims of philosophy Wang proposes for himself the achievement of a world view (BAP, p. 27), but he is aware that the complexity of the world and the massive accumulation of relevant information may make this aim unattainable in the foreseeable future (*ibid.*). He suggests, as a retrenchment, "the possibility of concentrating on some important and large aspects of the totality to look for a 'synthesis,' either as a preparation for the more remote end or as a self-contained finished product or preferably as both" (*ibid.*). In the pursuit of synthesis he maintains an attitude of openness, which he ascribes in part to his Chinese cultural background: "one of my primary concerns in studying philosophy has been to consolidate and apply the range of beliefs on which reasonable people agree, rather than to engage in detailed debates or present bold views that stimulate and provoke responses" (LJ, p. 21). Under the influence of John Rawls he found an affinity, surprising at first sight, between moral philosophy and philosophy of mathematics:

> It seems to me useful to study both moral philosophy and philosophy of mathematics with a view to narrowing the range of disagreement within them. Doing so provides us with complementary illustrations of ways of linking persistent philosophical controversies more closely to what we know—in contrast to the usual mutual criticism limited to a high level of generality. (LJ, p. 344)

That the urbanity and rejection of contentiousness contained in these attitudes of Wang toward the cultivation of philosophy are compatible, and indeed continuous, with a quite strong formulation of principles of philosophical methodology is the thesis of the remainder of this paper.

3. Factualism and its consequences in Wang's early philosophy

"Substantial factualism" as a method was proposed by Wang in an early philosophical book (FMP) but was never disowned in his later works in spite of modifications and refinements.

> The underlying belief is in the overwhelming importance of existing knowledge for philosophy. We know more about what we know than how we know what we know. We know relatively better what we believe than what the ultimate justifications for our beliefs are. (FMP, p. 1)

This manifesto is clearly a rejection of a large part of the modern tradition of epistemology, both in the rationalist version of Cartesian doubt and the empiricist version of Locke's "plain historical method." Although Wang is seriously interested in the history of philosophy and is fully cognizant of his historical evaluations, he presents his case for substantial factualism first and most extensively in the context of foundations of mathematics and only later generalizes to other parts of philosophy.

Substantial factualism in mathematics is the method of respectful attention to the actual practice of mathematicians, which almost always leads to agreement regarding the truth or falsity of mathematical propositions, even when disagreements persist concerning foundations.

> In the 1950s I was struck by the impression that what the different schools on the philosophy of mathematics take to be the range of mathematical truths form a spectrum, ordered more or less by a linear relation of containment that exhibits a step-by-step expansion. (LJ, p. 214)

> Our historical experience shows that such extensions have not produced irresolvable contradictions or even lesser difficulties. On the contrary, on the whole and in the long run,

strong agreement tends to prevail among practicing mathematicians, if not always over the issue of importance, at
least over that of correctness. (*op. cit.*, p. 215)

One of the principal motivations for the more abstemious philosophies of mathematics, such as finitism, intuitionism, and predicativism (the disallowance of quantification over sets of natural numbers) is the alleged risk resulting from the relaxation of prohibi-
tions. Factualism is an antidote to this motivation, for it points
out that serious contradictions in mathematics historically are rare
and are almost always resolved by further research (*e.g.*, paradoxes
concerning infinite series, resolved by Cauchy, Weierstrass, and others). Evidently, factualism is not only a compilation of historical
data but also an extrapolation and an attitude: an induction that
the remarkable successes of the past will continue in the future, together with a bet, of the kind characterized by Isaac Levi [1967] as
"gambling with truth," that garnering the riches of a strongly based
mathematics upon the assumption of its consistency has greater intellectual utility than the marginal increase of security provided by
a weak basis. Wang expresses this gamble dramatically by quoting,
apparently with approbation, Gödel's conjecture, "If set theory is
inconsistent, then elementary number theory is also inconsistent"
(quoted in LJ, p. 216) and commenting:

> It is not excluded, although it is not likely, given our accu-
> mulated experience, that some new paradoxes will be found
> in set theory or even in classical analysis. If that should
> happen, there would be alternative possibilities: we might
> either find some convincing local explanation or trace the
> trouble back to the initial big jump to the infinite. (*ibid.*)

Wang believes that factualism as a method has some implications
concerning the epistemology of mathematics and possibly about
its ontology, though he seems to be uncertain regarding the latter.
By endorsing an intellectual bet on rich mathematics, factualism
suggests an epistemology that is strong enough to account for this
richness (BAP, pp. 13-22). He also cites with approbation Gödel's
criticism of the view that true mathematical propositions are analytic in the narrow sense of being tautological:

In this sense ... even the theory of integers is demonstrably non-analytic, provided that one requires of the rules of elimination that they allow one actually to carry out the elimination in a finite number of steps in each case.[2]

He equally approves of Gödel's alternative explication of 'analytic':

A proposition is called analytic if it holds, owing to the meaning of the concepts occurring in it, where this meaning may perhaps be undefinable (i.e. irreducible to anything more fundamental). In this sense the axioms and theorems of mathematics, set theory, and logic all are analytic, but need not, as a result, be "void of content."[3]

Wang also agrees that judgments of analyticity in this sense require an intellectual capacity which was generally denied by empiricists, a faculty of conceptual intuition. In fact, Wang's substantial factualism endorses the recognition of this faculty, because an unbiased, as contrasted with a dogmatic, empiricism reveals introspective experience of relations among concepts that is as compelling as sensations and perceptions.[4] Gödel goes further by maintaining a Platonistic ontology as the complement of this epistemology. But Wang seems unsettled about the need for a Platonistic ontology. He has a very favorable section on Bernays, who was committed to mathematical objectivity and nevertheless concludes from the set-theoretical paradoxes that we must give up "absolute Platonism" (LJ, p. 339).

Wang's factualism differs sharply from Quine's extreme empiricism, according to which the distinction between mathematical and natural scientific knowledge is only a matter of degree. Wang regards this view of empiricism as imposing an implausible dogma upon factualism (BAP, p. 161). He says, "for example, set theory seems pretty well self-contained" (*op. cit.*, p. 160). This argument could be strengthened, and possibly it was in a passage of Wang that

---

[2] Gödel [1944], p. 150, quoted in BAP, p. 14.

[3] BAP, p. 14, a modified quotation from Gödel [1944], p. 151.

[4] Charles Parsons commented in correspondence, "Surely the factualist position doesn't justify the claim of 'introspective experience of relations among concepts that is as compelling as sensations and perceptions.' That's a phenomenological claim that someone who accepts mathematics and science isn't committed to, although it may be part of an *explanation* of mathematical knowledge."

I missed, by pointing out that a disproof of a mathematical proposition of great generality is achieved by a counter-example, which is an act of mathematical intuition just as that which was supposed to establish the general proposition; whereas a counter-example to a proposed general law of physics requires both a physical measurement and the psychological process of reading the physical measuring apparatus.

Finally, Wang recognizes two almost immediate methodological consequences of factualism. One is fallibilism: even though the richness of mathematical knowledge supports the epistemological thesis that human beings are endowed with rational intuition, there is no reason to ascribe infallibility to this intuition. Anomalies have been found in mathematical practice, though reflection by experts has so far corrected them; but it is only an induction—possibly erroneous—that this correction will always be achieved in the future. The other consequence is the appropriateness of the hypothetico-deductive method to mathematics. Some set theoretical axioms, for example, are not intuitively evident but imply many propositions which are "verifiable" in the sense of satisfying the intuitions of experienced mathematicians and do not imply any counter-intuitive propositions. Wang thus agrees with Gödel, and also with Russell in one of his phases, in admitting hypothetico-deductive reasoning in mathematics as much as in theoretical physics (RKG, p. 180; LJ, pp. 244-5). Regarding both fallibilism and the hypothetico-deductive method, however, there are some differences of interpretation between Wang and Gödel, which will be presented in the following section.

There are several other domains besides mathematics in which Hao Wang asserts the validity and utility of factualism as a method, even though the precision of implications of mathematical factualism are not claimed to be equaled elsewhere. Examples he cites in the early work FMP are certain parts of the natural sciences: classical physics and Mendelian genetics (FMP, p. 1). Soon afterwards he adds, "In its general conception it is not intended to apply exclusively to knowledge in the exact sciences. We are also interested in less exact knowledge and less clearly separated-out gross facts" (*op. cit.*, p. 2). It is striking that Gödel and Wittgenstein, opposed in many respects but both fascinating to Wang, agreed in their emphasis on the philosophical centrality of the gross facts of ordinary life. He quotes Gödel as saying "Everyday knowledge, when analyzed into components, is more relevant [than science] in giving data for philosophy" (LJ, p. 297), and he also quotes a similar passage from Wittgenstein's *On Certainty* (LJ, p. 366).

There are two domains which at first sight are surprising as loci for the application of factualism: the history of philosophy (LJ, pp. 326-327) and moral judgments (*op. cit.*, p. 368). In both of these the prevalence of conflicting opinions seems *prima facie* to constitute counterexamples to the thesis of factualism. But Wang finds in both cases that further reflection mitigates the appearance of unresolvable conflict. There is a residue of Hegel in his discernment of intellectual progress as a result of philosophical disputation (LJ, pp. 308-310), and he says, with evident admiration that "Leibniz was outspokenly proud of his own capacity to select and synthesize salient features of alternative views" (LJ, p. 327). As to moral judgments, he is greatly impressed with Rawls's conception of "reflective equilibrium" (*op. cit.*, pp. 348-353), which he proposes to extend to other domains than moral judgment. The inclusion of history of philosophy and of moral judgments as domains for the application of substantial factualism shows quite clearly that this methodology overlaps with a dialectic method—an overlap which is explicit in the latter part of LJ, his final book, but which probably was implicit even in his earlier work. This suggestion will be explored in more detail in §5.

4. Discussions between Gödel and Wang on methodology

In reporting a conversation with Gödel in 1971 on the manuscript of his FMP, Wang reveals a striking partial consensus and partial divergence between the two.

> Gödel ... saw factualism as a philosophical method and said that in his younger days he had taken something like it to be the right approach to philosophy. He did not deny that the method is of value, but said that it had intrinsic limitations and should be used in conjunction with Husserl's phenomenological method and with lessons from historical philosophy—especially in the pursuit of fundamental philosophy. (LJ, p. 326)

Since Gödel's philosophy is one step away from Hao Wang's, which is the concern of this paper, and Husserl's is two steps away, only a brief comment will be made here about Gödel's attraction to phenomenology. As noted in §3, Gödel maintained that a true mathematical proposition is analytic "owing to the meaning of the concepts occurring in it" (BAP, p. 14), and a crucial part of his epistemology was the attribution of an intellectual intuition to human beings. Husserl's phenomenological method was an intuitive examination of meanings, which he distinguishes from investigations of empirical introspectionist psychology. It is not clear that Gödel accepted this distinction of Husserl (see items 5.3.27 and 5.3.28 of quotations from Gödel in LJ, p. 169), but Husserl's promise of providing a systematic method for the clarification of meanings appealed to him (see his paraphrase of Husserl in item 5.1.3 on p. 158 of LJ) and, at least for a while, seemed to him feasible. There were several spectacular instances of clarification of meanings in mathematics in an apparently unique way that weighed heavily in this inclination of Gödel's towards Husserl's method of 'intuiting essences': especially Turing's characterization of computability, of which he said, "We had not perceived the sharp concept of mechanical procedures before Turing, who brought us to the right perspective. And then we do perceive the sharp concept" (*op. cit.*, p. 232). Gödel says cautiously that "Husserl speaks of constituting mathematical objects but what is contained in his published work on this matter is merely programmatic" (LJ, p. 256), but then immediately follows this caution with "Phenomenological investigation[5] of the constitution of mathematical objects is of fundamental importance for the foundations of mathematics."

Gödel's preoccupation with conceptual analysis extended beyond mathematics to philosophy (including theology), psychology, physics, and practical life (LJ, chapter 9). Regarding the first of these Wang says,

> The main task of philosophy as he saw it was (1) to determine its primitive concepts, and (2) to analyze or perceive or understand these concepts well enough to discover the

---

[5][The published text reads "investigations," but that would make the sentence ungrammatical. Gödel could equally have meant "investigations ... are," but the difference is hardly significant.—Eds.]

principal axioms about them, so as to "do for metaphysics as much as Newton did for physics." (*op. cit.*, p. 288)

Gödel expressed some disappointment, however, concerning the success of Husserl's investigations. When Wang asked for a specific impressive achievement of Husserl, "he mentioned Husserl's work on time, but added that the important part had been lost" (*op. cit.*, p. 320). Wang had the impression that, regardless of some disillusionment with Husserl, Gödel remained optimistic about the prospects of conceptual analysis of fundamental concepts beyond mathematics. "He believes there is a sharp concept corresponding to our vague intuitive concept of time—but we have not yet found the right perspective for perceiving it clearly" (*op. cit.*, p. 321). Wang is skeptical even of the modified and programmatic version of phenomenology which Gödel seems to fall back upon. Chapter 10 of LJ is largely devoted to presenting the deviation between him and Gödel, beginning with the following condensation:

> In my opinion, the use of lessons from the history of philosophy is an integral part of factualism; but phenomenology is a special type of reductionism, and factualism is an attempt to avoid the pitfalls of all types of reductionism. For Gödel, however, phenomenology is a way to carry out Platonism, which is, he believes, the right view; even though Platonism, too, is in a sense reductive, but acceptable because it is a "reduction" to the universal. It seems to me that Gödel's strong form of Platonism is at the center of the several major points on which I am unable to agree with him. (*op. cit.*, pp. 326-327)

Gödel's optimism about the power of intuitive conceptual analysis is largely an extrapolation from his own successes in mathematics and logic, but Wang claims that Gödel has somewhat misrepresented his own working methods:

> His own spectacular work was obtained otherwise: by applying thoroughly the familiar method of digesting what is known and persisting, from an appropriate reasonable perspective, and with exceptional acumen, in the effort to see and select from a wide range of connections. Undoubtedly, careful reflection pointing in the direction of Husserl's

method played a part too — but only in combination with thinking based on material other than the act of thinking itself. (*op. cit.*, p. 332)

It is precisely this other material that Wang thought was unavailable when Gödel attempted an exact science of metaphysics, possibly along the lines of a monadology, which was appealing to Gödel. "It is unnecessary for me to say that I am unable to see, on the basis of what we know today, how such an ideal is likely to be realized in the future. From the perspective of factualism, I believe we know too little to give us any promising guidance in the pursuit of this grand project" (*op. cit.*, p. 332). More generally, Wang finds both a positive and a negative side to factualism: the recognition of knowledge in oneself when it is present and recognition of absence of knowledge when it is absent. (Cf. the beautiful epigraph of Confucius heading ch. 10 of LJ, p. 323.)

It is striking that Wang appeals to factualism when he expresses skepticism of Gödel's optimism regarding *a priori* procedures in almost the same way as when he expresses skepticism about Quine's subsumption of logic and mathematics under the empirical. Wang says, in fact, that "the distinction between the a priori and the empirical, like that between the innate (or the hereditary) and the acquired, points to something fundamental that is hard to delineate in any unambiguous manner" (*op. cit.*, p. 332). Wang wishes to bypass the difficult problem facing both Gödel and Quine of establishing the delineation between the *a priori* and the empirical by using instead "the system of universally available and acceptable general concepts and beliefs" (*op. cit.*, p. 333). In contrast with Gödel's suggestion that "The Newtonian scheme was obtained a priori to some extent" (quotation 9.2.37, LJ, p. 303), he asserts that "From the perspective of factualism, we are entitled to appeal to concepts and beliefs grounded on our gross experience" (LJ, p. 333).

The deviation between Gödel and Wang on the employment of the axiomatic method (or equivalently, the hypothetico-deductive method, as I understand Wang's usage) is crucial and rather subtle. Gödel occasionally appealed to this method in mathematics, most famously in his discussion of the continuum hypothesis (RKG, p. 180; LJ, pp. 248-253) and he continually emphasized its importance in physics and philosophy (e.g., LJ, pp. 334-335). Wang notes that Gödel's conception of an axiom system is more liberal than that of formal system (LJ, p. 334) and specifically that he regards Newton's mechanics to be an axiom system. Both Gödel and Wang agree that axioms often have to be revised in the light of evaluations of their consequences. Where they differ is on the prospect of transcending the hypothetico-deductive method to achieve knowledge that is more definitive and more intuitive. The hypothetico-deductive method was for Gödel an instrument, perhaps for human beings an indispensable one, to expedite the clarification of fundamental concepts. He remained optimistic about the capability of human beings to achieve this clarification, as indicated by the passage cited above about the concept of time (LJ, p. 321). Wang, by contrast, seemed to accept the indispensability of the hypothetico-deductive method as a consequence of the constitution of human cognitive faculties: "As our knowledge and intuition develop, we may find new axioms or revise old ones. Sometimes we need new information from the outside, such as observations and experiments of physics and biology." (LJ, p. 335)

I have maintained in several publications ([1993a] and [2002]) that the achievement of a coherent philosophy requires the "closing of the circle" of epistemology and metaphysics, with the hypothetico-deductive method as a bridge between these two great components. This thesis, the first part of which is derived from Aristotle's Posterior Analytics and Metaphysics, does not appear to be articulated explicitly either by Gödel or Wang, but it seems to be adumbrated in the foregoing quotation from Wang.

5. Wang on dialectic

Although Wang was skeptical about Gödel's attraction to Husserl, he shared with Gödel some reservations about the sufficiency of factualism as a philosophical method and in his late works made a number of suggestions for supplementing and refining factualism. Many of these suggestions are recommendations for employing dialectic. Here is a list of varieties of dialectic that Wang considers at one point or another.

(a) In complex matters dialectic is an antidote to one-sidedness.

132

It has been suggested that the work of Marx and Wittgenstein points to the way of developing a social theory which avoids the one-sided pitfalls of subjectivism and objectivism ... I would have thought that any systematic study of the human situation would have to find a proper mixture of the subjective and the objective. In fact, I would prefer to speak of the general and the particular and their interplay. In this regard, I believe that dialectics is the much abused "method" (or rather style) or guidance toward a judicious organization of the complex interrelations (say, between the general and the particular, between subject and object, subject and subject, etc.) But what is needed of dialectics is a knowing how rather than a knowing that, a skill to use it rather than a doctrine to expound it. That is probably why Marx never wrote any extended treatment of dialectics. (BAP, p. 208)

(b) In a discussion of the use of history of philosophy Wang says:

Our ideas develop through a complex dialectic of what we learn from the outside versus our own thoughts ... For instance, the interplay of my wishes with circumstances led me to certain views and to a familiarity with certain parts of human knowledge, including the work of certain philosophers. ... By reflecting on my agreements and disagreements with them, I have come to understand better my own views, as well as some of the reasons why people disagree in philosophy. (LJ, p. 327)

(c) Wang sometimes identifies dialectic with some generalized sense of "logic":

The vague but suggestive word dialectic is commonly used to describe the interaction of contradictory or opposite forces that lead to a higher and more unified stage in some processes. Traditionally, dialectic is closely associated with logic. Indeed, throughout the Middle Ages the word dialectica designated what we now call logic. For Hegel, logic is the science of the dialectical process—the continual unification of opposites in the complex relation of parts to a whole. (LJ, p. 16)

That this passage is more than a summary of some of the history of philosophy is shown by the following remarks on the role of logic in philosophical methodology in the final chapter of LJ:

> One heuristic guide to the development of logic as metaphilosophy is the ideal of being able to see alternative philosophies as complementary. (LJ, p. 363)

> In a sense, logic is the instrument for singling out the definite and conclusive parts of our thoughts. And it is tempting to suggest that, within philosophy, such parts all belong to logic.... I am inclined to think of the range of logic as consisting of all those concepts and beliefs which are universally acceptable on the basis of our common general experience—without having to depend on any special contingent experience. (LJ, p. 364)

> Logic as an activity of thought deals with the interplay, or the dialectic, between belief and action, the known and the unknown, form and content, or the formal and the intuitive. For this purpose, it is useful to select and isolate from what is taken to be known a universal part which may be seen, from a suitably mature perspective, to remain fixed and which can therefore serve as instrument throughout all particular instances of the interplay. It seems natural to view such a universal part as the content of logic. (*op. cit.*, p. 366)

(d) The last in the litany of dialectical opposites just quoted, the dialectic of "the formal and the intuitive"—sometimes also characterized as "the *dialectic* of intuition and idealization" (*op. cit.*, p. 210)—deserves special attention, since it plays an essential role in Wang's, and also in Gödel's, philosophy of mathematics. Both recognize that the axioms proposed for a body of logic and mathematics are often less intuitive than propositions which are derived from the axioms. Consequently, even though suitable axioms are indispensable for systematization, some of them are confirmed by the intuitions into the truth of their consequences. In other words, the dialectic of the formal and the intuitive is an adaptation to mathematics of the hypothetico-deductive method of the empirical sciences, as Gödel asserted in his philosophical article [1947] on

Cantor's continuum hypothesis. The dialectic of the formal and the intuitive also goes in another direction, since formalization can clarify the relations among concepts and hence sharpen intuition. This direction of the dialectic is what we should expect from Gödel, in view of his attraction by Husserl's phenomenological method. But Wang (despite his reservations about Gödel's 'rational optimism') also agrees at least partially:

> The process by which we arrive at the formal systems F, H, and N from our intuitive understanding of finitary, intuitionistic and classical number theory obviously involves the dialectic of the intuitive and the formal, of intuition and idealization. What is more remarkable is that, once we have the formal systems, we are able to see precise relations between them which enable us to gain a clear intuitive grasp of some of the relations between the original concepts. (LJ, p. 219)

(e) A variant of the foregoing sense of dialectical opposites is this:

> We may go from small sets to large sets by the dialectic of intuition and idealization, intuitive overview and thought, the subjective and the objective, knowledge and existence. (LJ, p. 254)

This litany is suggestive but vague. It indicates that Wang (and probably Gödel also) do not think of a simple schematic application of the hypothetico-deductive method to foundations of mathematics. A detailed examination of Wang's philosophy of mathematics is essential for a precise understanding of dialectic within this domain, and that I must leave to experts. The remainder of my discussion is largely conditional: if it is assumed that Wang's philosophical methodology is successful within the domain of mathematics, how well does he extrapolate it, or how well can it in principle be extrapolated, to the grand project of formulating a coherent and systematic world view?

(f) There are several further statements about philosophical methodology, some of them dignified with title of "principles," which Wang does not, so far as I have noticed, characterize explicitly as "dialectical," but surely deserve at least to be characterized as "crypto-dialectical." Of these, the most striking are his endorsements of elements of John Rawls's theory of justice, a part of moral and political theory that is very remote from the domain of Wang's expertise—and all the more striking, since Wang is generally opposed to reductionism and to hierarchies in philosophy, recommending instead attention to the different methodological requirements of different disciplines. He asserts, however,

> It seems to me useful to study both moral philosophy and the philosophy of mathematics with a view to narrowing the range of disagreement within them. Doing so provides us with complementary illustrations of ways of linking persistent philosophical controversies to what we know—in contrast to mutual criticisms limited to a high level of generality. (*op. cit.*, p. 344)

I cannot summarize adequately all that Wang extracts from Rawls in section 10.3 of LJ and all the less can I do justice to Rawls's own detailed and subtle arguments. But Rawls's concept of "reflective equilibrium" deserves special attention, because it epitomizes a plausibly achievable goal of moral dialectic and also suggests some ways in which the dialectic can be effectively carried out:

> Rawls's notion of reflective equilibrium aptly captures a fundamental component of methodology which many of us have groped after.... For Rawls, considered judgments are those given when conditions are favorable to the exercise of our powers of reason. We view some judgments as fixed points, judgments we never expect to withdraw. We would like to make our own judgments both more consistent with one another and more in line with the considered judgments of others, without resorting to coercion. For this purpose, each of us strives for judgments and conceptions in full reflective equilibrium; that is, an equilibrium that is both wide—in the sense that it has been reached after consideration of alternative views—and general—in the sense that the same conception is affirmed in everyone's considered judgments. Thus, full reflective equilibrium can serve as a basis for public justification; which is nonfoundationalist in the following sense: no specific kind of considered judgment, no particular level of generality, is thought to carry the whole weight of public justification. (LJ, pp. 349-350)

Elsewhere he summarizes the idea of reflective equilibrium in a way
that clearly points to general philosophical applicability:

> In every domain we make considered judgments at all levels
> of generality. In order to arrive at some sort of systemati-
> zation of our considered judgments, we reflect at each stage
> on the relations both between such judgments and between
> them and our intuition. In this process we continually mod-
> ify our considered judgments with a view to finding, eventu-
> ally, a set of considered judgments in reflective equilibrium.
> (*op. cit.*, p. 348)

(g) In a section entitled "Alternative Philosophies and Logic as
Metaphilosophy" (LJ, pp. 361 ff), Wang proposes two principles
which are crypto-dialectical in that they serve to guide the dialectic process. One is the Principle of Limited Mergeability:

> Of two conflicting beliefs, if there be one to which all or al-
> most all who (1) understand or (2) are aware of all the rea-
> sons for both of them give a decided preference, irrespective
> of their other views, that is the better justified belief. (LJ,
> p. 369)

The other is the *Principle of Presumed Innocence*:

> What we suppose we know is presumed to be true unless
> proved otherwise... I think of this principle as an antidote
> or a corrective measure to what I take to be an excessive
> concern with local or uniform clarity and certainty. For ex-
> ample, ... I need not pay too much attention to skepticism.
> (LJ, p. 71)

The second of these seems to be a restatement of factualism, which
was central in Wang's earlier philosophy and retained with modifi-
cation in his later thought; indeed, it is nothing more than factualism adapted to the interplay of dialectic. The first of these two
principles is somewhat disturbing, because it poses a dilemma of
choosing between continuing the dialectic, in order to determine
who understands and who knows all the relevant reasons, or terminating the dialectic by some criterion (that may well be challenged)
for identifying the people who do understand and do know reasons. Indeed, having expressed some misgivings in the preceding

paragraph about a specific one of Wang's methodological principles, I wish to ask questions (similar to considerations raised by Parsons ([1998], pp. 20-21)) which may trouble readers of my summary of Wang's methodology and indeed of his entire exposition: even if one grants the attractiveness of everything he says about tolerance, open-mindedness, avoiding one-sidedness by extracting valuable ideas from *prima facie* conflicting philosophies, shunning excessive skepticism, subdividing problems for more manageable treatment, identifying conflicts that are due to diverse meanings of the same words, using logic as an instrument of clear thinking, etc., isn't he essentially asserting commonplaces and therefore failing to make a substantial contribution to philosophical methodology? Doesn't the medication he prescribes go down too easily to be effective? I think the answer to both questions is negative, but some explanation would be desirable.

One of the epigraphs of LJ is a quotation from Hegel's *Greater Logic*:

> *It is only after profounder acquaintance with the other sciences that logic ceases to be for subjective spirit a merely abstract universal and reveals itself as the universal which embraces within itself the wealth of the particular.* (LJ, Preface, p. ix)

Whatever Hegel meant by this passage and however Wang exactly construed it, it does suggest the lesson that the dialectic cannot be properly carried out without attention to detail. In the preceding paragraphs Wang does indeed make some methodological statements which he intends to be applicable in all domains of philosophy, but he is steadily anti-hierarchical and anti-reductionist and therefore takes these statements to be schematic, inapplicable in any specific domain without detailed attention to the particularities of that domain. Wang personally entertained a desire to achieve a coherent world view and regarded the professional discipline of philosophy as having this aim, but as noticed earlier he was troubled by the massive accumulation of philosophically relevant human knowledge, far beyond the capability of any individual to master. Nevertheless, he was able to give two instances of enriching his general methodological statements with attention to the details of a discipline, and these are suggestive of at least an attitude

138

and frame of mind that would be successful in other disciplines. One instance, of course, is the discipline of philosophy of mathematics, which he discusses in detail in large parts of four books and in many research papers. Being an expert in a field as highly structured as logic and mathematics is, of course, no surrogate for expertise in other fields, but it provides an exemplary experience of what constitutes expertise; and since logic explicitly or tacitly underlies the reasoning employed in other disciplines, Wang's own expertise must have given him, with some justification, confidence when he came to examine argumentation elsewhere. The other discipline in which his general methodological principles are enriched by attention to details is moral philosophy, in which he does not claim personal expertise but relies, with good judgment, upon the expertise of John Rawls (with admixtures from sources as diverse as Confucius and Marx). The difference between moral philosophy and philosophy of mathematics regarding amenability to precise formulation, entanglement with emotions, susceptibility to cultural influences, requirements of worldly experience, demands deep psychological insight, etc. is valuable for the grand project of aiming at a world view: the two domains impose such different demands upon the particularization of general methodological principles by attention to details that perhaps they may jointly prepare the aspiring philosopher for achieving this particularization in domains which lie at intermediate points between mathematical and moral philosophy in the entire philosophical spectrum. A striking example is given in §6, in a passage by Wang concerning Marx's philosophy of history.

Another element which raises Wang's practice of dialectic above the level of commonplaces is the pervasive cultivation of open-mindedness, which was remarked in §2 of this paper. The abstract dialectic pattern of synthesizing opposites can become mechanical and sterile (in a way that my former teacher Richard McKeon characterized as "the freezing of the dialectic"), and genuine open-mindedness—whether innate or cultural or acquired by education—is an antidote against this danger. At many points in Wang's exposition there is a glow of open-mindedness: in his effort to transform oppositions among schools of mathematical philosophy into the recognition of a spectrum of options, in his sturdy (not entirely convincing) effort to exhibit a complementarity of Gödel's and

Wittgenstein's philosophies of mathematics (LJ, pp. 327-332), and in the delightful juxtapositions of passages from Chinese and Western philosophers. Perhaps one of the lessons that Wang teaches by example, without preaching, is that the dialectic works only when it is practiced with good character.

Finally, the dialectic method as practiced by Wang derives strength and effectiveness from its union with factualism, which he retained, with elaboration, from his discussion of philosophical method in FMP. In paragraph (g) above, it was noted that the crypto-dialectical Principle of Presumed Innocence is a restatement of factualism. Wang never ceased to be astonished at the massive coherence and depth of mathematics, the natural sciences, and certain parts of the knowledge of ordinary life (especially the mastery of language). The citation of an appropriate part of this massive knowledge in a dialectical analysis of philosophical possibilities can, and does in fact, open new pathways of analysis that are not accessible to armchair practitioners of dialectic. (A few examples among many are the theory of inertial motion, the age of the earth, relativistic space-time structure, and the molecular theory of genetics.) And apart from opening new pathways of dialectical analysis, attention to relevant scientific discoveries often lends detail and weight to one of several competing possibilities in a philosophical debate. As noted in §3, factualism is an extrapolation and a gamble, wagering that future scientific discoveries will not jettison—although they may modify and refine—propositions in the contemporary canon. At several points Wang quotes Gödel's characterization of himself as a "rationalistic optimist." It seems accurate to apply to Wang the characterization "empirical optimist," and then to add that such optimism is reasonable in the light of the astonishing body of working human knowledge. Epistemologists with a decision theoretical orientation (including myself) have also justified empirical optimism as a strategy that is not sure to succeed, but will do so if any epistemological strategy will succeed. (See Shimony [1993b], p. 298). I have not, however, found any passage in Wang's methodological writings explicitly subscribing to this 'pragmatic' argumentation.

I conclude this discussion of the strength of Wang's dialectic method with a note on a curious omission. Among the various lists provided by Wang of dialectical theses and antitheses, I have nowhere found an explicit mention of the relation between epistemology and ontology—although it is possible that the opposition of the subjective and the objective, which he mentions several times, is broad enough to subsume this relation. Epistemology investigates the assessment of human claims to knowledge or justified belief, and ontology investigates the basic entities of the world and their properties. Each of these great branches of philosophy is obviously relevant to the other, and, in my opinion, must be investigated in tandem (despite the recurrence in the history of philosophy of programs which aim at the establishment of a reliable epistemology prior to, or dispensing entirely with, any ontological commitments). The coherent meshing of an epistemology with a metaphysics—which has been called "closing the circle" (Shimony [1993a], p. 21 and [2002], p. 311)—is a reasonable condition of adequacy for a systematic philosophy. Furthermore, by explicitly aiming at this closure, one can enrich the philosophical dialectic by drawing freely upon the best scientific knowledge we have of the place of human beings in the cosmos, of their evolutionary antecedents and achievements, and of the workings of their sensory and cognitive apparatus. In other words, a naturalistic epistemology is part of the program of systematic philosophy, and it offers much greater detail and perspective than epistemologies which are methodologically more abstemious—such as introspective and phenomenological and analytic epistemologies. Now it seems to me that Wang is sympathetic to the idea of "closing the circle" and tacitly appeals to it in his dialectical practice, in spite of his reluctance in Ch. 7 of LJ, mentioned in §3 above, to agree explicitly with Gödel's commitment to an ontology of mathematical entities.

Another aspect of "closing the circle" that Wang addresses is the constitution of the human mind that is required for it to have insight into relations among mathematical objects, if one supposes that such objects exist. Wang cites Gödel's preoccupation with this problem, saying

> Gödel seems to believe both that the mind is more complex than the brain and that the brain and the human body could not have been formed as a matter of fact entirely by the action of the forces stipulated by such laws as those of physics and evolution. (LJ, p. 193)

But then Wang comments, "Of course, the desire to find 'holistic laws' has been repeatedly expressed by many people. As we know, however, no definite advance has been achieved so far in this quest" (*ibid.*).

This fragment of a discussion between Gödel and Wang may provide some insight into Wang's abstention from an explicit endorsement of a dialectic of epistemology and ontology. He might have thought that we know too little scientifically about relevant matters to engage fruitfully in such a dialectic. Several pages of BAP (pp. 202-204) confirm this reading of Wang, for they express skepticism about the amount of light that empirical investigations in cognitive psychology, psychoanalysis, and developmental psychology have thrown upon interesting epistemological problems. In defense of the program of "closing the circle" against this skepticism I shall briefly mention two considerations. One is that there is an exemplary case of illumination of the epistemology of ordinary life by studies of neurology and psychology: that is, the study of processing of optical data in the eye and the brain whereby reliable inferences are habitually made in ordinary circumstances about middle-sized physical objects of the environment (see Gregory [1966] and Neisser [1967] for valuable summaries). The other is that as regards scientific investigations of the higher achievements of human thought empirical psychology is indeed disappointing, but the science is young, and Wang—of all people—should be aware of the need for patience.

I shall summarize this survey of Wang's work on philosophical methodology—laudatory but with some reservations—by stating my overall reaction: that Wang is one of the few authors I have read since my student days who have given me the exhilarating feeling of new intellectual horizons in philosophy.

6. Wang on Marx and Mao

Some of Wang's autobiographical remarks concern his early training in Chinese philosophy (*e.g.*, LJ, pp. 122-126), which he never ceased to take seriously in spite of his later immersion in western philosophy. He quotes several times from Confucius, and notes with obvious approbation the latter's goal of a stable and contented society (LJ, p. 122). However the various influences on Wang's thought combined, an outcome for several years following the summer of 1972 was a "will to believe" (his phrase, *op. cit.*, p. 124) in a Maoist version of Marxism. He then wrote:

> From 1977 to 1979, a less distorted picture of the actual situation in China was gradually revealed to me through personal conversations and published accounts of what had happened. Slowly I began to realize that my belief about what was happening was fundamentally incorrect. (*ibid.*)

A less personal statement of the evolution of his thinking, which incidentally shows the application of his philosophical methodology to social philosophy, asserts that the Marxist theory of historical determinism

> includes a plausible general view of history and an acute analysis of the capitalist society. The crucial and probably most speculative point is probably the identification of the dynamic force of human history with the proletariat, which is taken to be a more or less homogeneous class of individuals who share a number of special properties, ideally suited for the grand role. The situation is analogous to proposing a solution to a very complex set of equations ...

> For many years Marx and his followers believed that the working class in several Western European countries actually or potentially satisfied these conditions to a large extent. Historical experience showed that this was not quite the case and we do not possess a simple set of equations for the (unknown) dynamic force that admits a presently determinable group of individuals as its solution. It is familiar today that nationalism remains a strong force and the working class has become less homogeneous and less uniformly impoverished. (RKG, pp. 257-258)

The hopefulness expressed at the beginning of this passage exhibits a certain basis of factualism (especially regarding Marx's acute analysis of capitalist society), as well as a mathematician's paraphrase of Marxist determinism and a large emotional ingredient. The conclusion of the passage is pure factualism, acknowledging the unwelcome contingencies of history. It would be hard to find in the entire literature on Marxism a passage which so concisely expresses so much thought.

References

Cited writings of Hao Wang:

1958a. Eighty years of foundational studies. *Dialectica* 12, 466-497.

1974a. *From Mathematics to Philosophy.* London: Routledge & Kegan Paul.

1985a. *Beyond Analytic Philosophy. Doing Justice to What We Know.* Cambridge, Mass.: MIT Press.

1987a. *Reflections on Kurt Gödel.* Cambridge, Mass.: MIT Press.

1996a. *A Logical Journey. From Gödel to Philosophy.* Cambridge, Mass.: MIT Press.

Other cited writings:

Gödel, Kurt, 1944. *Russell's mathematical logic.* In Paul Arthur Schilpp (ed.), *The Philosophy of Bertrand Russell,* pp. 125-153. Evanston: Northwestern University. Reprinted in Gödel [1990].

Gödel, Kurt, 1947. What is Cantor's continuum problem? *American Mathematical Monthly* 54, 515-525. Reprinted in Gödel [1990], pp. 176-187.

Gödel, Kurt, 1990. *Collected Works,* volume II: *Publications 1938-1974.* Edited by Solomon Feferman, John W. Dawson, Jr., Stephen C. Kleene, Gregory H. Moore, Robert M. Solovay, and Jean van Heijenoort. New York and Oxford: Oxford University Press.

Gödel, Kurt, 2003. *Collected Works,* volume V: *Correspondence H-Z.* Edited by Solomon Feferman, John W. Dawson, Jr., Warren Goldfarb, Charles Parsons, and Wilfried Sieg. Oxford: Clarendon Press.

Gregory, R. L., 1966. *Eye and Brain.* New York: McGraw-Hill.

Koehler, Eckehart, 1999. Review of Wang [1996a]. In Jan Wolenski and Eckehart Köhler (eds.), *Alfred Tarski and the Vienna Circle,* pp. 312-318. Vienna Circle Institute Yearbook, 6. Dordrecht: Kluwer.

Levi, Isaac, 1967. *Gambling with Truth*. New York: Knopf.

Neisser, Ulrich, 1976. *Cognition and Reality*. San Francisco: Freeman.

Parsons, Charles, 1998. Hao Wang as philosopher and interpreter of Gödel. *Philosophia Mathematica* (3) 6, 3-24.

Shimony, Abner, 1993a. Reality, causality, and closing the circle. In Shimony, *Search for a Naturalistic World View*, vol. I, *Scientific Method and Epistemology*, pp. 21-61. Cambridge University Press.

Shimony, Abner, 1993b. Reconsiderations on inductive inference. In *ibid.*, pp. 274-300.

Shimony, Abner, 2002. Some intellectual obligations of epistemological naturalism. In David B. Malament (ed.), *Reading Natural Philosophy: Essays in the History and Philosophy of Science and Mathematics*, pp. 297-313. Chicago and LaSalle, Ill.: Open Court.

Wang and Wittgenstein[1]

Juliet Floyd

*I would like to begin and end with a classification
of what philosophy has to attend to. The guiding
principle is, I believe, to do justice to what we
know, what we believe, and how we feel.*

Hao Wang, *Beyond Analytic Philosophy*

1. Introduction

No account of Hao Wang's philosophy can be complete without
discussion of his serious engagement with Wittgenstein. This began in the period 1953-1958, immediately following the publication
of *Philosophical Investigations* [1953] and *Remarks on the Foundations of Mathematics* [1956], and culminated in a second phase of
engagement from 1981 to 1995. It closed in the final chapter, "Alternative philosophies as complementary," of Wang's posthumously
published *A Logical Journey: From Gödel to Philosophy* ([1996a],
cited hereafter as LJ). Wang's work on Wittgenstein thus preceded
the discussions he had with Gödel (starting from 1967), and at least
partly shaped his articulations of Gödel's philosophy. Unlike Gödel,
Kreisel, and Bernays—with whom he worked on logic and philosophy and whose readings and criticisms of Wittgenstein Wang took
seriously—Wang was inclined increasingly over time to claim that
Wittgenstein, despite certain limitations of his approach, had made
fundamental and constructive contributions to philosophy.

The purpose of this essay is to characterize what Wang thought
those contributions were, to say something about why he held

[1] I owe numerous debts to the editors for their great patience and help in
bettering this essay and in producing this volume of essays. Thanks are also due
for comments on a late draft by my colleague Tian Cao and Juliette Kennedy.

Wittgenstein in such esteem, and to evaluate Wang's contribution to our understanding of Wittgenstein. This will preclude discussing Wang's most important contributions to logic, mathematics, philosophy, and their history. Fortunately others have broached discussion of these, especially Wang's discussions of and with Gödel.[2] I hope to be able to provide a snapshot of his engagement with a particular philosopher that will convey something of Wang's own philosophical ambitions and temperament. Though his influence on philosophy proper is not widely recognized today—partly because of the unorthodoxy of his rejection of much of the analytic philosophy of his day, and partly because of difficulties internal to his own thought—Wang's role in shaping more than one generation's understanding of the fundamental problems in logic and their history through the 1950s was significant, and not as widely acknowledged as it should be.[3] Perhaps more important, philosophy was for Wang himself the most central and significant subject, so that if we wish to measure his own sense of his accomplishments, this part of his work cannot be ignored.

Before I begin, certain qualifications are in order. Wang, to his credit, was never a "Wittgensteinian" in the sense of being a single-minded *devoté*. He always denied that he was an expert scholar of Wittgenstein, and was even proud of the fact that in his essay [1961b], an assembly and analysis of passages from *Remarks on the Foundations of Mathematics*, he never once mentions the philosopher's name.[4] He admired Wittgenstein mainly for the challenges

---

[2]See Parsons's essay "Hao Wang" in this volume, as well as his [1996] and [1998] and Shieh [2000].

[3]Parsons has described his brief studies with Wang in the mid 1950s. (See his [1998], note 10, quoted and amplified in note 2 to the Preface in this volume.) In conversation Burton Dreben stressed with me more than once the important role Wang played for him and others interested in logic at Harvard beginning in 1946. Wang educated students in basic proof theory, an area in which Quine was neither especially focused nor especially adept. Because Wang taught at Harvard and Oxford, the cumulative impact of his teaching on the dissemination of logic was significant. Perhaps as important, Wang supported study of the subject as internal to philosophy proper. Hide Ishiguro has stressed to me how supportive Wang was of Michael Dummett during his early years teaching at Oxford, when logic was not a very popular subject among philosophers there.

[4]Wang died five years before the release of the electronic version of Wittgenstein's *Nachlass* (2000; published as [2003]), and in most of his writings had

he posed and the suggestions that he made, but only against a wider backdrop of Wang's own philosophical projects and his interests in the history of philosophy, mathematics, and science. There were parts of Wittgenstein's philosophy that Wang positively rejected, as well as parts he viewed as too one-sided, even if valuable. Certainly he felt that Wittgenstein's philosophy, while of fundamental importance, was unclearly formulated and deserving of better articulation in light of its alternatives. All of his treatments of Wittgenstein take place from within a broader philosophical project, and most of his discussions of Wittgenstein are, implicitly or explicitly, comparative. This illustrates his approach in philosophy more generally. Readers of Wang's accounts of Gödel's philosophical thoughts should, I think, bear this in mind. Wang was not reading either Wittgenstein or Gödel neutrally, but charitably and critically, in terms of his own philosophical ideas. He was doing philosophy, not merely describing it.

In what follows I shall focus on a number of themes pertaining to Wittgenstein that extend through both phases of Wang's evolution, tying these to Wang's own writings. I shall be illustrative, rather than fully explanatory or directed at details, as it is impossible to attempt anything like an exhaustive characterization of Wang's enormous corpus of philosophical writing, even limiting myself to the role of Wittgenstein. The suitability of this approach may be questioned, of course. But my sense is that directly critical assessments of Wang's particular arguments, while valuable, may miss the wide forest of his views for the trees. For it was not argumentation, but reflection, discernment, and synthesis of knowledge, that were his *fortes*. I shall focus on reconstructing what I take to be some of the most central insights and challenges for Wang's readers.

These include Wang's notions of "perspicuousness," "factualism,"

---

to work with memoirs of Wittgenstein's students, rather than the biographies by McGuinness [1988] and Monk [1990]. But his knowledge of the corpus in the early 1990s was remarkably well-informed, and I doubt that viewing the e-version would have altered in any significant ways his understanding of Wittgenstein any more than did the biographies. The remarks from the *Nachlass* that would have surely most interested him are those on Gödel's theorem that were unknown until after 2000, and those on Turing's use of the diagonal method. Floyd [1995] was directly inspired by conversations with Wang; cf. Floyd [2001]; Floyd [forthcoming a].

148

"conceptualism," and "intuition," as well as his idea of the "dialectic" of the formal and the intuitive. These were the notions used by Wang to present his own view of the nature of knowledge, including philosophical knowledge.

2. Wang's reading of Wittgenstein in context

2.1. *Wang and Other Readings of Wittgenstein.* Wang belonged to a gifted generation of philosophers who constructively yet critically engaged with the first and most intensive wave of Wittgenstein's reception in the mid-1950s: Anscombe, Cavell, Dummett, Feyerabend, and, though less explicitly in his published writings, Rawls.[5] Like them, Wang offers a fresh and critical approach to Wittgenstein, different both from those of Oxford ordinary language analysis and from Kantian readings of Wittgenstein common in the 1960s and 1970s. The philosophical standards against which Wang's reactions to Wittgenstein are to be measured are thus of a high caliber, and his analyses, especially of Wittgenstein's remarks on mathematics, are to be counted as lasting, even if neither definitive nor ultimately correct.

Like these contemporaries, Wang took Wittgenstein's anti-empiricism and anti-reductive conceptual pluralism, as well as his concern with probing the character of objectivity and agreement as embodied in actual practice, as central philosophical concerns. Like them, Wang did not see ordinary, everyday language as offering an ultimate subject of theorizing or a constraint on speculation, but rather, at best, a touchstone or challenge for reflection.[6] He did not

---

[5] At Harvard the initial reception of Wittgenstein was significantly shaped by Rogers Albritton and Burton Dreben, at least in the classroom, and not by Wang. Until the early 1970s, however (by which time Wang had left Harvard), Dreben mainly taught logic, with an occasional seminar on *Remarks on the Foundations of Mathematics* (Wang had begun teaching Dreben logic while he was still an undergraduate, and provided him with an introduction to Paul Bernays before 1950). Albritton (and Cavell) were the major influences on the study of the later Wittgenstein's significance for epistemology. Hilary Putnam took a logic course with Wang at Harvard and later discussed logic with him while at Oxford, but Wang's philosophy seems to have had little impact on him.

[6] In fact, Wang was more passionately against "linguistic philosophy" than any of these philosophers, even favorably quoting Gellner's [1959] irreverent

take the concept of truth to be exhaustively characterized by any one theory, or even a central topic or approach.[7] And he was not concerned to dismiss ontology.[8] He was always loath to dismiss a philosophical perspective with the term "nonsense."

Though he wrote on Hume's problem of induction, Wang did not have much interest in drawing philosophical lessons from a confrontation with general forms of skepticism ([1950c], FMP; cf. LJ, p. 371). His readings of Wittgenstein reflect this, differing not only from those of Cavell—who makes skepticism of central importance to Wittgenstein—but also from the later readings of Kripke and Wright, who take one of Wittgenstein's central contributions to have been a new form of skepticism about following a rule.[9] The contrasts with the latter, widely known reading are instructive, both for what they reveal about Wang's general philosophy and for how he read Wittgenstein.

Wang did not take Wittgenstein's form of constructivism to be based upon a general concept of rule-following, logical instantiation or necessitation, or conceptual grasp. There are of course more than a few ways to criticize a reductive idea that "meaning is use," and rule-following skepticism is one of them. But another way is *via* a sophisticated form of conceptualism in which grasp of concepts

---

anthropological comparison between ordinary language philosophy and a secular, established religion for gentlemanliness ([1974a] (cited hereafter as FMP), p. 393). He explicitly praised Rawls's *A Theory of Justice* [1971] for its willingness to go beyond the "special kind of piecemeal linguistic or conceptual analysis" Wang took to have wrongly dominated Anglo-American philosophy in the 1960s (LJ, p. 326).

[7]Wang emphasized here the importance of Wittgenstein's remarks on truth in his [1980], p. 75: "Philosophy is not a choice between different 'theories'. It is wrong to say that there is any one theory of truth. For truth is not a concept." However, Wang certainly did think that truth is a concept, and that Gödel, for example, had shown us at least some of its essence.

[8]Nor did he take Wittgenstein to have been preoccupied with such a dismissal. He welcomed, for example, von Wright's pointing out to him that the term "metaphysical" was an erroneous transcription of the word "metaphorical" in early editions of *Culture and Value*: Wittgenstein remarked that "there is no religious denomination in which the misuse of metaphorical [JF: not "metaphysical"] expressions has been responsible for so much sin as it has in mathematics." Cf. LJ, p. 181 and Wittgenstein [2003] MS 106, p. 58 (1929), correctly transcribed and translated in Wittgenstein [1998], p. 3e.

[9]During the conversations I had with Wang between 1990 and 1995, I often asked him to comment on these readings, but he resisted discussion of them.

150

is understood in terms of something other than the purely deductive model. This latter approach was Wang's. He always took Wittgenstein to have been a conventionalist about at least part of the content of mathematical knowledge, but this did not exhaust his understanding of what Wittgenstein had to say.

By now there is much in Wang's writings on Wittgenstein that seems dated, but in other respects his remarks on Wittgenstein are still relevant. Wang would certainly have shunned fictionalist accounts of mathematics, including those inspired by the idea— arguably Wittgensteinian—that mathematics belongs to our artifactual capacity to invent and represent, rather than attaching directly to actual truth and knowing. Rather than ascribing to Wittgenstein a response-dependent, anti-realist, or assertion-conditional account of meaning, Wang took him to be investigating phenomena of certainty that figure in logical and mathematical objectivity at the basis. Wang would have welcomed recent work in philosophy of mathematics in which visual elements of diagramming are made central to the theory of reasoning, computer modeling is regarded as of fundamental philosophical interest, and in which, more generally, the philosophy of mathematical practice is given center stage.

2.2. *History and Philosophical Method.* Fact and fiction, concept and object, actuality and possibility, were for Wang notions best placed within the frame of logic and the theory of knowledge. But though he took them to be structured by the historical fact of idealization in science and the ubiquitousness of its methods of mathematization, including formalization of proof, Wang resisted the reduction of philosophical methods to logico-formal methods throughout his life. This was an important element of the affinity he found with Wittgenstein and with Gödel and explains the disaffection he had with the main figures of the analytic tradition in twentieth century philosophy, including Quine, who was one of his most important teachers.

Unlike Wittgenstein, Wang used historical perspective as a weapon in mounting his criticisms.[10] This expressed part of his

---

[10]In conversation Wang explicitly rejected a remark Wittgenstein reportedly made in the early 1930s in reaction to Broad ([1980], p. 74-75):

If philosophy were a matter of choice between rival theories, then

factualist ideal. Perhaps for this reason his writings are distinctive in containing a wealth of unusually measured, insightful, informative, and objective—as opposed to dramatically narrated—history. Here is a contrast with Feyerabend, as also with Kuhn. Like them, under the influence of Wittgenstein, Wang resisted the positivist's ideals of empiricism and the use of formalized languages to solve fundamental philosophical problems. Yet Wang urged resistance from a variety of different points of view, and with full knowledge of the contributions of mathematical logic to philosophy and to mathematics, as well as a serious understanding of the history of mathematics proper. Typically his arguments are laced with specific examples drawn from the history of mathematics, logic, and philosophy, and express no general account of science or scientific method.

Early on Wang adduced limitative results on decision procedures: "the quest for an ideal language is probably futile. The problem of formalization is rather to construct suitable artificial languages to meet individual problems" ([1955e], p. 236). Historically, he pointed out, the distinction between "artificial" and "natural" languages is a matter of degree, familiarity, and culture.[11] He also highlighted the role of mathematical notation (not merely formal logical notation) as something crucial to, even if not exhaustive of, our grasp of thought, especially in allowing us "perspicuous" grasp of massive details ([1961b]). Later he emphasized that "the philosophically more central and more difficult task is to grasp the right ideas intuitively; how far they can be or how well they are

---

it would be sound to teach it historically. But if it is not, then it is a fault to teach it historically, because it is quite unnecessary; we can tackle the subject direct, without any need to consider history.

[11]Noting the long development of the spoken Chinese language over time, he stressed that although each alteration of the language seemed at the time of its introduction "natural," if one had tried to introduce the changes all at once, one would have been attempting to make a kind of revolution, and this would probably have failed. But "on the other hand, when an artificial language meets existing urgent problems, it will soon get generally accepted and be no longer considered artificial," so that "it may be more to the point if we compare artificial languages with Utopian projects" ([1955e], pp. 236-37). Wang's picture here is that ideal aspirations may realize themselves over time, at least to some extent, but never fully and in every detail.

formalized is an auxiliary secondary consideration, which is admittedly very helpful sometimes" ([1985a], p. 122).[12] In the history of philosophy, Wang believed in presenting not only results and principles, but biographical facts as well, thus using individuals, not archetypes, as his narrative frame. This expressed a conviction that philosophy should be relevant to individual human life and human experience, not merely to the facts, but also a belief that biographical and/or cultural facts about a philosopher may be relevant to an assessment of his or her philosophy.

In his later writings on Wittgenstein and Gödel, Wang found congenial what he took to be their firm respect for facts, including facts of everyday life and experience in their own lives, however idiosyncratic ([1991?], p. 23). This makes reading his accounts of their philosophies difficult. Ordinary biographies of these figures are easier to take in for those interested in the sweep of their intellectual lives, and critical conceptual analyses are by contrast the norm for philosophers. In Wang's writings there is a mixing of the genres, with his own sense of history and philosophy overlain on top. This does not mean that these writings fail to contain a wealth of detail from which many different readers can learn. But it does imply that one should bear in mind Wang's own philosophical ideas in assessing their value. It also explains why Wang's writings are not, and may well never be, tremendously influential. He had great ambitions and a discerning eye for fundamental work, but he found in the end that a convincing synthesis and articulation of his own perspective eluded him.

2.3. *Objectivity Before Objecthood.* The philosophical frame surrounding Wang's philosophy is difficult to characterize, for it is multifaceted and schematic. In the end, it offers no more—and no less—than a "flexible" yet comprehensive framework for thinking about philosophy, both in its scientific and its literary articulations ([1991?], p. vi). Certain general things may, however, be said. Unlike Dummett, whom he knew well and admired, Wang did not take the theory of meaning to be a fundamental branch of philosophy. He stressed the importance of alternative characterizations of the relation between language and thought to twentieth century philosophy, focusing on the multiplicity and vicissitudes of languages, not

---

[12]This work will be cited hereafter as BAP.

their unity. Facts as we know them must, on his view, form a direct object of reflection quite apart from their linguistic expression, placing non-trivial constraints on philosophical theorizing.

The primacy of facts over objects was something Wang admired in Wittgenstein's *Tractatus*, and tended to encapsulate in sayings reminiscent of one long associated with Kreisel: "objectivity before objecthood" ([1991b], pp. 260ff.; [1991?], p. 71). The idea was to take objectivity to require only a bifurcation of propositions into true and false (by the law of excluded middle), thereby leaving open how best to articulate the nature of objects. Wang's "substantial factualism," as he called it in FMP, was intended to develop this idea in a systematic way, providing an alternative to the forms of linguistic philosophy he resisted. He proposed calling it "anthropocentric magnifactualism," thereby tying it to his earlier reading of Wittgenstein as an advocate of "anthropologism" (FMP, p. 1). But Wang knew that Wittgenstein did not regard the notion of "fact" as of much use in elucidating the nature of mathematics.

Wang's "conceptualism," indebted to Dedekind and Gödel, is hardly to be classified as Wittgensteinian, except in the amorphousness of its edges and the absence of a general account of concepthood. It turns away from the methods by means of which Wittgenstein drew distinctions between factual and conceptual investigations. Yet Wang believed that the very idea of conceptual knowledge implies a contrast and a connection with technical or combinatorial skill, and he proposed late in life to characterize his own perspective with the term "connectivism," rather than "conceptualism," a Wittgensteinian sounding revision ([1991?], pp. 262-3). He was never sympathetic to Wittgenstein's critical remarks about Dedekind, regarding as too one-sided their persistent griping about the dominance of the extensional point of view.[13] Yet Wang understood that what Wittgenstein opposed was what Wang himself

---

[13]It is interesting to note that Bernays [1959] is relatively charitable to these particular remarks of Wittgenstein, though he regards them as ultimately unsatisfactory, as it seems one must. Bernays takes Wittgenstein to be resisting the mixing up of intensional and extensional approaches, an "applicable" criticism, according to Bernays, in certain versions of the Dedekind theory of numbers "that create a stronger character of the procedure than is actually achieved." Bernays thinks Wittgenstein's considerations are of potential value in combatting the kind of dogmatism that sometimes accompanies reductions— an idea Wang would pick up on and develop in his own writing on Wittgenstein.

154

regarded as a kind of dogmatism, exemplified by philosophers like Quine. Wang certainly regarded reasoning by means of infinitary objects as a primary given that forces us to accept certain principles in the philosophy of mathematical science, as Wittgenstein did not. And he felt, quite understandably, that Gödel was right to suggest that Wittgenstein sinned against his own philosophical stance in coming close to denying the existence of facts about sets ([1991?], p. 35). Wang certainly felt that Wittgenstein too often disregarded a principle he should have embraced fully: that "the requirement of leaving things alone demands also a fuller appreciation of the alternative views to one's own favorite" (ibid.).[14] And he believed that Wittgenstein saw "true philosophy as insulated from mathematics," as he could not (LJ, p. 19). Wang endorsed, rather than merely questioned, Dedekindean views that model mathematical objectivity on law-preserving extensions of concepts, and he seemed prepared to take such views more or less at face value, as Wittgenstein did not.

On the question of philosophical approaches to the concept of objectivity, however, a more nuanced difference may be discerned. Wang appreciated the positive role of formalization in carrying forward the logicist project, but appreciated Wittgenstein's criticisms of a reductive attitude toward formal proofs. He understood the limitations of a vague appeal to "law governedness" in explaining the objectivity of structures, and he was not inclined to take second-order logic to have provided us with a philosophically sufficient basis for the theory of concepts. Insofar as the Dedekindean ideal of rigor is taken to be primarily that of the axiomatic method, Wang appreciated Wittgenstein's willingness to question the application of that ideal across the board in philosophy, as he believed Gödel had not. Gödel, Wang believed, was overly optimistic about the use of the axiomatic method in philosophy—an assumption Wang felt the Vienna Circle had shared (BAP, p. 104). But "the essential immunity of mathematics to the contingent vicissitudes of language cannot be shared by philosophy," according to Wang (LJ, pp. 210-11).

---

[14] It is also true, however, that in discussion with me in the early 1990s Wang was particularly puzzled about the status of Wittgenstein's writings on mathematics in the early 1940s. He sensed that Wittgenstein had invested a great deal of effort here, and felt these writings had not yet been well enough understood. That judgment still stands today, I believe.

Thus Wittgenstein served in Wang's mind as a useful corrective to overly optimistic, uncritical over-extensions or oversimplifications of method. In general, proof and axiomatization are in Wang's view important tools for distilling explicit acknowledgment of principles which can then be assessed and discussed, but they are never means for unearthing self-evident, unrevisable facts. Their application is generally limited to mathematics and to purely logical questions. But the foundations of logic: these lie in broader features of language and thought not reducible to mathematical science and unlikely to be resolved in a definitive way.

2.4. *Logic and Foundations.* What made Wang's engagement with Wittgenstein most distinctive, then, is that he reached his interpretive conclusions by grappling directly with Wittgenstein's remarks on mathematics and logic, and then attempted to draw general lessons for philosophy as a whole. Wittgenstein's philosophy could potentially be used, he believed, to make significant contributions to the foundations of logic at a basic level in a way that could contribute to a philosophical program of potentially wide methodological significance.[15] He seems to have felt that Wittgenstein could help him bridge the gap between his technically specialized knowledge of mathematics and logic and overarching, general philosophy, and he believed that this was where the value of Wittgenstein's writings on mathematics were to be found. His focus on Wittgenstein was thus narrower than that of any of these readers of Wittgenstein I have mentioned, though his ambitions were as broad.

Throughout his philosophy, and certainly in his readings of Wittgenstein, Wang's inspiration was the *Grundlagenstreit* in the foundations of mathematics. He was deeply inspired by Paul Bernays's call in [1935], in the face of the bitter controversies between Hilbert and Brouwer, for an informed and broad-minded approach to foundational issues in mathematics and philosophy, one

---

[15]von Wright and Hintikka spring to mind as similar readers in this respect. But the idea of solving philosophical problems directly via logic was anathema to Wang. He explicitly differed with von Wright's views about the history of logic, finding them too narrow and disengaged from the history of mathematics, as well as overly pessimistic about the potential relationship between logic and philosophy in the twenty-first century. This is clear from comments Wang made on the manuscript of von Wright [1994] in a letter of 10 September 1991 that I discussed with him.

that would avoid exaggerated claims about "crises" in foundations, exercise minimal partisanship and maximal reasonableness of approach, and portray the history of logic and mathematics as capable of proceeding, ultimately, in harmony and security despite a plurality of approaches.[16] He followed Bernays's idea that broad talk should be replaced with a careful and informed mathematical exploration of the consequences and possibilities for a variety of logical and philosophical alternatives.

Nevertheless, Wang was not inclined to regard mathematics or logic as capable of settling philosophical disagreements on their own: logic in particular had philosophical foundations, and just here was where he located Wittgenstein's specific contributions. He admired Wittgenstein's ways of exploring the elements of fundamental philosophical choices and moves and emphasizing the complexity and variety of ways in which thought may be seen to be expressed in language. He believed that a "neutral viewpoint" that adopted a "detached position" from certain philosophical controversies would be productive (LJ, p. 214). When explanations give out, the philosopher should describe, and not explain or defend. The best that may be done is to present a range of alternatives and arrange them, if possible, in a synoptic, step-by-step manner. This, Wang felt, would allow for the defense of what he called "open-minded, stepwise Platonism" ([1991?], p. 54), and avoid overreactions to apparent paradoxes. But such arrangement itself belongs to what may broadly be conceived of as logic.

2.5. *Logic as Metaphilosophy.* Increasingly over time Wang became explicit that he had devoted himself to articulating an approach, nowadays deemed (perhaps tendentiously) "quietist," in which the totality of possible approaches would be carefully canvassed, distilling what might be deemed best in each. He did not mean by this that all philosophical problems could be reduced to matters of language use. In fact, he considered Quine's proposal that existence be analyzed by way of the uses of pronouns in particular languages to be "harmful and dangerous," on the ground that, given how we

---

[16]Bernays says at the beginning of [1935] that the mathematical sciences are growing in harmony and security, though he does not claim this about logic or foundations. Wang, however, conceived of logic as at least ideally playing an adjudicative, mediating, and harmonizing foundational role in philosophy.

now speak and think, we are in "no good position to anticipate the form of all confusions ahead of time" ([1991?], pp. 142-3).

Yet while recognizably drawing upon and reacting to Wittgenstein, Wang's version of "substantial factualism" is complex and hardly reducible to any simple-minded descriptivist view (for more see Parsons [1998]). He did not believe that as the sense of crisis in foundations passed, philosophical disputes would fall away as unnecessary residue or merely pragmatic dross, to be replaced with strictly mathematical, cooperative work. Instead, he believed that the nature of philosophical dispute should itself be normatively reassessed in light of the history and philosophy of logic and mathematics, as well as the broader history of philosophy.[17]

Like Bernays, he stressed that Wittgenstein too often rejected speculation. And he was steadfastly critical of the trend in analytic philosophy, shaped by Carnap, that erected or rejected philosophical distinctions primarily by formal means, grounding choice of framework on vaguely enunciated, pragmatic appeals to the scientific enterprise as a whole or to a general holism about evidential support. Already in his [1958a] Wang complained about the trend toward "piecemeal exercises" in philosophy, having in mind not only the exercises of ordinary language philosophers, but also those of technically minded analytic philosophers who were, in his mind, too reductive and sanguine about the usefulness of methods of formalization in philosophy and in mathematics ([1958a], p. 468).

Wang regarded resistance to dogmatism and partisanship in philosophy, whether it was theoretical or practical, not only as essential to the health of the subject, but essential for overcoming the political and humanitarian challenges bequeathed by philosophies of the twentieth century. He had in mind, of course, not only disasters perpetuated in the name of philosophy in the West, but also in China. Top-down, totalizing philosophies that fail to respond to individual perspectives and feelings, philosophies that ignore the practical need to reach agreement on collective action through dialogue, philosophies that claim infallible insight, denying the possibility of a reasonable range of disagreement: these were all anathema to

---

[17]In conversation Wang told me that one reason he had found his professorship at Harvard unsatisfying is that he was appointed in the Department of Applied Mathematics, and not in Philosophy, where he felt the bulk of his efforts and his contributions had been made.

158

Wang. Much of his philosophical motivation came from a wish to devise a philosophy responsive to these concerns.

These attitudes partly explain, I believe, why Wang was attracted to Wittgenstein's later philosophical writing, designed as it was to investigate and to question the philosophical "must."[18] Wittgenstein's later methods and style of writing were attractive to Wang in their way of asking for philosophical responses from readers one by one, from the bottom up, without dogmatism. As he saw it, they invite the idea of trying to reflect upon a whole range of possible individual responses to philosophy, and, in fact, to the experience of human life as such. Wang at one point defined philosophy as "an attempt to attain a perspicuous view of (all of) human experience" ([1991?], p. 21), which he equated with "the quest for (comprehensive) perspicuous objectivity" ([1991?], p. v), or a worldview.[19] Wittgenstein's remarks thus fulfilled a certain methodological ideal to which Wang was attracted: state all reasons for and against the available positions and then let the reader make up her own view on a subject ([1991?], p. 144).

Wang himself did not, however, shun comprehensive and systematic theory in philosophy, and he faulted Wittgenstein for doing so. In the end, however, Wang himself did not achieve a formulation of a philosophy that satisfied him, or could be easily taken in. He was perhaps for this reason attracted at the end of his life to the later writings of Rawls. These Wang explicitly adduced in the epilogue to his final book as an especially valuable exemplar of philosophizing with the right aims and methods (LJ, chapter 10).

Like Rawls, Wang regarded the idea of a social contract understood in terms of coordination and pure convention, a mere *modus vivendi*, as inadequate. A deeper measure of agreement based on a normative theory was needed, but one that would be based on the here and now, that is, on how we as individuals respond to the idea of the legitimacy of an historically given structure or sys-

---

[18]Wang goes so far as to say that a major task in the philosophy of mathematics is to either justify or to explain away the common feeling that elementary truths about small integers are intrinsically obvious and convincing ([1991?], p. 155).

[19]At [1991?], p. 44, Wang wrote, "It is, I believe, perfectly reasonable to view Gödel's conceptual realism and Wittgenstein's earlier and later philosophies as alternative attempts to give a perspicuous view of the human experience (and therewith of the world)."

tem. In his early work, Wang tended to treat this as something like "the sociological fact" that a result is accepted ([1961b], p. 329; [1987b], pp. 88ff.), but as he progressed, objective agreement became the quest of philosophy proper, and in this context something more elusive, more normative, more open-ended, and difficult to articulate.

What the later Wang emphasized in the work of the later Rawls is, in fact, more Wittgensteinian than Kantian. For Wang was steadfastly against the idea of utopian thinking where it is unwilling to see itself shaped by the friction of everyday life ([1955e], pp. 235-6; [1991?], pp. 154-155). Wang rightly took Rawls's distinction between treating a theory of justice as a comprehensive or metaphysical doctrine and treating it as a political conception forwarded from within the ongoing, constructed framework of "public reason" to be of fundamental methodological importance, a signal contribution to the history of philosophy as a whole, and not merely to political theory. At the same time, he agreed with Rawls that clarity will not leave philosophy, or the world, as it is, but potentially alter it ([1991?], p. 32).

In his own writing, however, Wang did not proceed systematically, as did Rawls, but synoptically and intuitively, as he took Wittgenstein to have done (cf. Parsons [1998], p. 20). He took one of Wittgenstein's observations more literally to heart than his contemporaries: in the face of certain fundamental questions and distinctions, where explanations end, supplement pure philosophy with descriptions and observations. Wang's efforts in this direction lack the literary sophistication of Cavell's and Feyerabend's writing and not infrequently tend toward being overly descriptive or biographizing, especially in his later writings on Wittgenstein and Gödel. Yet Wang's assemblies of reminders are not lacking in stylistic sophistication, and they are never trivial. For they are everywhere informed by his extensive knowledge of history and forwarded in service of his wider philosophical point of view.

2.6. *Wang's evolution: Gödel and Wittgenstein.* Wang's work on Wittgenstein may be organized into two phases, forming a pair of parentheses around his engagement with Gödel. In the first (typified by his [1955e], [1958a] and especially [1961b][20]), indebted to

---

[20] Wang mentions Wittgenstein rather dismissively in his earliest essay

Bernays, his central concern was to make sense of Wittgenstein's remarks on mathematics by analyzing them in relation to Bernays's [1935] idea of a hierarchical, five-fold way of regarding the distinction between constructivist and non-constructivist positions: anthropologism (Wittgenstein's alleged position, "strict finitism" in the language of Kreisel [1958]), finitism, intuitionism, predicativism, and Platonism. He used Wittgenstein to structure his understanding of the notion of "effectiveness" and his understanding of what he took, throughout his life, to be a fundamental "dialectic" between the "formal" and the "intuitive."

The second phase (after 1981) took place after his discussions with Gödel and his formulation of factualism. Here Wang aimed to draw from Wittgenstein's later writings on the nature of logic and philosophy methodological lessons for philosophy as a whole.[21] At this juncture the comparison between Wittgenstein's philosophy and Gödel's was of central interest, and a characterization of their contributions to twentieth century philosophy the overarching aim. Gödel and Wittgenstein seemed to Wang to provide examples of true depth in philosophy, complementary alternatives to the then prevailing empiricism, emphasis on formal methods, conventionalism, and holism that he associated with Carnap and Quine, and vehemently disliked.

Wang tended to think of the contrast between Wittgenstein and Gödel as exemplifying the wider and important thematic dialectic between the pull of the "intuitive" and the pull of the "formal" or "idealized" or "conceptualized," a dialectic he applied to account for the history of science, East and West, and the history of philosophy.[22] The contrast also showed the importance of a general

---

([1945]/[2005], p. 139), a review of Russell's *Inquiry into Meaning and Truth*. He grouped him with Bergson as a philosopher holding, self-contradictorily, that "there is some knowledge that language cannot be used to express." Wang later came to soften his view of Wittgenstein by devising a distinction between the intuitive and the formal, but it is not clear how the paradox is overcome by this means. In the 1990s Wang was especially interested in discussing with me the then "new" interpretations of the *Tractatus* that aim to resist the kind of flatly contradictory reading he had once proposed.

[21]Chapter two of BAP works out ideas about Wittgenstein's early philosophy as a kind of "digression"; Wang announces a plan to write more substantially on Wittgenstein in a later place (p. 75).

[22]Wang for example thought of Western medicine as more "formal" and conceptualized, and Chinese medicine as more "intuitive" ([1984a], p. 528).

conception of logic to philosophy. In his final writings, Wang drew on Wittgenstein's *On Certainty*, less as a response to skepticism than as a starting point for presenting his own "factualist" ideal in terms of the concept of perspicuousness and his own conception of "logic as metaphilosophy" (LJ, chapter 10), a comparative and adjudicative method Wang adopted as his own.

In the end Wang rejected two aspects of Wittgenstein's philosophy that were crucial to it: 1. Wittgenstein's attempt to work without a fixed structure of overarching epistemological distinctions, a general theory of the mind, or of logic; and 2. Wittgenstein's insistence on pursuing philosophy and/or logic through an investigation of grammar, or logical distinction-drawing, without relying directly on traditional, historically given systems of thought. Wang was always skeptical of the idea that a focus on how we use language to express thoughts might yield philosophical illumination. So, in contrast, he developed an overarching set of epistemological distinctions, highly indebted to the past, and attempted to treat the philosophical data in terms of the framework they provided. Even if he regarded his frame as "flexible" and merely schematic ([1991?], p. vii), even if he attempted to revise and critically revisit traditional philosophical categories, he took these as a given, and worked with them.[23]

Wang himself located his most fundamental philosophical differences with Wittgenstein at the same level as he located his most fundamental differences with Gödel: in the greater generality and richness of view he thought his own philosophy had achieved:

> My own perspective differs from theirs in what I see as a different conception that favors a sharper delineation of the several distinct kinds of human experience. For instance, within the frame of experiential objectivism I aim to give the distinguishing traits of our mathematical experience their due, and shun away from seeing it as representative of all experience or as a "degenerate" though pervasive kind of experience that forms a part of "grammar." In particular, I

[23]Hence the justness, I believe, of Dreben's opinion that Wang, although "constantly deeply attracted to Wittgenstein," never accepted the "full force" of Wittgenstein (Parsons [1998], p. 21n). Wang's insistence on retaining a notion of "intuition" at the basis of his epistemology is a striking instance of this. See section 5 below, and note 10.

take it to be a philosophically relevant and significant fact
that mathematics (and physics) are "intellectual sciences,"
in which the danger of confusion due to (the use of) language
and the irresolvability of conflicting interpretations are not
so serious as elsewhere (say in the humanities and in art).
([1991?], p. viii)

3. Anthropologism

Wang appreciated early on that Wittgenstein was not an intuition-
ist, but in fact a critic of intuitionism, rejecting as an incorrect
idealization of mathematical proof the constructivist constraints on
reasoning offered by Brouwer and his followers. Wang was indebted
on this point to Kreisel's and Bernays's critical reviews of *Remarks
on the Foundations of Mathematics*.[24] They took Wittgenstein to
have advocated an even more restrictive form of "strict finitism,"
in which the use of an induction scheme allowing for generaliza-
tions over all numbers is rejected in favor of what is "feasible."
Wang agreed and chose for this position a term that is a "bit more
colorful," namely "anthropologism" ([1958a], p. 474; [1961b]). In
Wang's [1958a] the notion of "anthropologism" as an approach to
the foundations of mathematics is first framed. This is a definite
step beyond Bernays [1935], who mentions that one might question
unfeasible recursions but doesn't suggest this as a possible foundational stance.

One sees in his choice of terminology Wang's search for a broader
philosophical perspective from which the commitments of such a
position might be appreciated (elsewhere he speaks of "ethnologi-
cal" perspectives, alluding to Wittgenstein). Bernays had used the
phrase "anthropomorphic" to describe, more critically than Wang
would have, Wittgenstein's reliance on the character of grammar
and the fact of actual agreement in mathematics in connection with
logic ([1959], p. 30). Wang applauded the conceptual pluralism im-

---

[24]See Kreisel [1958] and Bernays [1959]. As late as [1987b], pp. 85, 89-90,
Wang took himself to be "in essential agreement" with Bernays's review. Later
on Wang would maintain that Wittgenstein was closer to the terms of variable-
free finitism, associated with Skolem, and deny that Wittgenstein was a strict
finitist. An example is ([1991?], 87ff.), which was however removed from its
successor passages published in (LJ, pp. 212ff.).

plied by Wittgenstein's idea, and its focus on practical agreement, as Bernays did not. For Bernays, there was a threat of irrationalism here. This Wang did not believe. For him Wittgenstein's insistence on "intuitiveness" was in a sense arational, insofar as it grappled with that which escapes conceptualization by one means or another, and in another sense perfectly rational, since such residue's presence was evident.

Wang never accepted "anthropologism" as a correct point of view. He saw it only an illuminating perspective along the way, an unearthing of steps of idealization. But one of the fundamental contributions Wang saw early on in Wittgenstein's finitistic-sounding remarks on mathematics was their illuminating the character of our need for an abstractive "big jump" (in Gödel's terms) to the totality of all (finite) numbers. Even Hilbert's finitism and Brouwer's intuitionism had taken this big jump for granted.

"Anthropologism" is an interesting and distinctive view, both of mathematics and of Wittgenstein, and Wang explored its presuppositions and consequences at just the time he did his most important work in what would now be called computer science. Though he deemed anthropologism a position clearly too restrictive to be satisfactory as a comprehensive view, he felt it worthwhile to consider as an antidote to the one-sidedness of reductionist aims in logical analysis he associated with Carnap. The notions of "perspicuousness" and "feasibility" brought out the importance of practice and concreteness of understanding, as opposed to theory.

Thus, unlike Kreisel, Wang did not see Wittgenstein's remarks on mathematics as "the surprisingly insignificant product of a sparkling mind" (Kreisel [1958], p. 158). Moreover, he did not consider the later Wittgenstein to have tended toward irrationalism or anti-scientific dogmatism, as did Bernays and Gödel. Instead, Wang drew out the constructive philosophical significance of Wittgenstein's notion of perspicuousness, both for the philosophy of mathematics and logic and for philosophy as a whole, interpreting it in terms of a notion of humanly "feasible" actions or computations, i.e., those that can "actually be carried out and kept in mind" ([1958a], p. 474). Wang's contributions to our understanding of Wittgenstein's thought are in this respect more substantial and wide-ranging than Bernays's or Kreisel's were. He aimed to explore the practical and philosophical consequences of Wittgenstein's re-

164

marks by situating them within a wider philosophical frame, where
their attractiveness and commitments could be clearly understood.

Wang admired Wittgenstein's deflation of forms of logicism that
construe it as offering a complete semantic analysis and/or grounding of arithmetical knowledge. In fact he agreed that no systematic
theory of mathematical reasoning can ground, in any interestingly
fundamental sense, our everyday knowledge of elementary arithmetical facts. The key notion he used in forwarding this criticism
is that of "perspicuousness."[25] As early as his [1958a, pp. 469ff.] he
wrote:

> When a reduction gives the impression of being of profound
> philosophical interest, there is reason to suspect... some
> trickery. The talk of logical foundations is misleading at
> least on two accounts: it gives the impression that number
> theory and set theory do not provide their own foundations
> but we must look for foundations elsewhere, viz., in logic;
> it implies that the grand structure of mathematics would
> collapse unless we quickly replace the sand underneath by a
> solid foundation. Neither thought corresponds to the actual
> situation. Indeed, if we adopt the linear mode of thinking
> to proceed from the logical foundation to the mathematical
> superstructure, there is surely something glaringly circular
> in the mathematical treatment of mathematics itself which
> makes up mathematical logic...
>
> The basic circularity suggests that formalization rather than
> reduction is the appropriate method, since we are, in foun-
> dational studies, primarily interested in irreducible concepts.

Of interest here is the contrast Wang draws between "formalization" and "reduction," and his focus on "irreducible" foundational

---

[25]German terms in Wittgenstein associated with this notion are *Über-
sichtlichkeit*, *übersehbar*, *durchsichtig*, *überblickbar*, *anschaulich*, and *ein-
prägsam*. These terms are used to modify presentations, proofs, and models,
though differently in different contexts of discussion, especially in mathematics
and philosophy. Wang and I discussed the usual translations into English in the
early 1990s ("perspicuous," "synoptic," and "surveyable"), and he continued to
use these terms, especially "perspicuous," as central to his understanding, both
of Wittgenstein and of philosophy itself. The connections with understanding
("taking in") and mastery in a practical sense were of central importance to
Wang here, but also the epistemic "intuitive" element in the sense of commanding a "firm" grasp of a concept ([1991?], p. 124). See note 28.

concepts. Wang tended to stress a series of contrasts between "conceptual" ("scientific," "theoretical") and "technical" ("practical," "intuitive") knowledge, treating the contrasts themselves as forming either a continuum or a dialectical pairing. Wittgenstein was always placed on the "practical," "intuitive," "concrete" side of the pairing, implying that his philosophy is fundamentally flawed in failing to do justice to the need for speculation, systematicity, truth, and theory. This Wang certainly believed.[26]

It is striking that in this early period Wang grouped under the broad term "formalization" a very wide array of phenomena not usually classified with the term: mathematization, conceptualization, but also all verbalization of thought and idealization. "To put thoughts in words or to describe a particular experience involves formalization of intuition," he held; and "it is impossible to formalize without residue the complete intuition at the moment" ([1955e], p. 231). This breadth of usage, historically indebted to Brouwer and the intuitionists, turns also on the idea that formalization offers a "translation" rather than a "reduction" (cf. [1958a], p. 470). Here are crucial indices of Wang's self-conception as a philosopher. They mark points at which he departed from Wittgenstein, whose broad conception of language included, e.g., the particular samples and paradigms used ostensively in the teaching and use of language. For Wang, Wittgenstein's later investigations of the variety of ways in which we might conceive the expression of thought in language were investigations of "formalizations," but not critical explorations of the notions of "language," "residue," "intuition," or meaning as such, as most readers would hold. Wang took the "dialectic" of the formal and the intuitive, the theoretical and the practical, the idealized and the intuitive, to be crucial for Wittgenstein, and then he adopted these *Leitmotifs* his own (LJ, chapter 7.1).

And yet in his early, most explicitly Wittgenstein-influenced essay [1961b], perhaps his overall finest as a single piece, a complex treatment of perspicuousness emerges through analysis of actual logical and mathematical practice, including the type of applications of logic in computer proofs which Wang had himself made.[27]

---

[26] Wang [1987b] states that the "anthropic" element in Kant's and Wittgenstein's philosophies of mathematics does not allow for the concept of arithmetical truth (p. 90).

[27] Wang's [1960b] reports that all the theorems of propositional and predicate

166

The question whether and in what sense *Principia Mathematica* might be fully formalized became for Wang an actual question, and the program he devised to derive some of its theorems a vindication of the usefulness of Wittgenstein's perspective on mathematics, which Wang proposed dubbing "praximism." The essay [1961b] is a subtle, prescient and informative piece of philosophical writing, laced with examples and suggestions for further thought. Wang is often quoting from or directly alluding to remarks of Wittgenstein (never with citation) while critically reflecting and commenting on their merits and limitations, often by confronting them with examples. Especially impressive is Wang's deft distinction-drawing in thinking through the interplay between logic and mathematics. Set theory is classified as mathematics, logic as a matter of breaking proofs down into small steps; "surveyability" and "perspicuousness" attach to the concept of proof, but to proof in practice, and not as necessary or sufficient conditions. Wang is careful not to reduce the issue merely to a matter of length of proof, the reproducibility of signs, formal systems, or epistemic certainty, though each of these issues figures in his choice of examples.[28] He sees that the notion of "perspicuousness" should be tied to reflection on sense-making, value, significance, interest, and meaning, and that its use lies in resisting reductions, not forwarding a one-sided point of view. An idea that he first formulated here would remain an important part of Wang's philosophical conception for the rest of his life: in the techniques of mathematics, but not necessarily those of logic, the practical and the theoretical merge.[29] But accounting for this merg-

logic in *Principia Mathematica* were proved in under nine minutes by the program he wrote. In this sense the *Principia* was given a precise formalization. In the early 1990s Wang said to me in conversation that this indicated to him the essential lack of conceptual richness involved in the *Principia*'s formulation of the deductive elements of mathematics.

[28]Marion [2009, forthcoming] and Mühlhölzer [2006, 2010] discuss a variety of understandings of Wittgenstein's notions. It would be interesting to compare Wang's uses with theirs, for it is less precise in emphasizing neither the visual side of the idea (as in Marion) nor the copying, reproducing pictorial idea (as in Mühlhölzer). Floyd [2000], an essay partly indebted to discussions with Wang, offers a broader use of the notion, focusing on the analogy between the perspicuousness of proof in mathematics and the perspicuousness of a presentation in philosophy.

[29]In the concept of "effectiveness" or "computability" as analyzed by Turing, this merging was most clear, according to Wang (cf. LJ, pp. 372-73, and [1961b],

ing requires of the philosopher careful attention to purposes, aims, and the richness of ordinary mathematical experience. Without factoring these in, misleading and dogmatic statements will and do abound. When Wang focuses on the distinction between calculation and experiment, he is taking calculation to be crucial to the logical aspect of proof, but not holding that all mathematics is calculation, or even that the heart of mathematics is calculation. The logician's idea that definitions are mere abbreviations; that decidability should be construed in terms of purely formal, explicit logical operations; that the possibility of correct interpretations of arithmetic requires the impossibility of incorrect interpretations of logic or of set theory; that the emergence of contradictions and/or lack of sharp boundaries with a concept immediately indicate a risk of conceptual incoherence it is obvious how to interpret—these ideas are given their due by Wang, but criticized. The point of emphasizing the "perspicuousness" of proofs in mathematics, then, was to avoid one-sidedness, not to promote a single point of view. It was to show how dogmatic and/or overextended the interpretation of a logician's perspective can be, but not to dismiss that perspective altogether.

Among the most important of the ideas Wang always rejected in Wittgenstein's later writings on mathematics are i) those voiced in Wittgenstein's harsh and dismissive remarks on Dedekind and set theory, with their suggestion that these parts of mathematics should be dismissed or restricted in favor of a more limited fragment based on a fully intensionalist view; ii) the idea that philosophical clarification "leaves everything as it is," dealing only with linguistic puzzles and rejecting speculation; iii) the idea that the notion of "convention" could be applied to account for at least a part of the fundamental content of mathematics; and iv) Wittgenstein's unwillingness to work with any notion of "intuition." Yet it is significant that even these rejected ideas were taken by Wang to incorporate elements of truth, and so aspects of Wittgenstein's thought that were worthy of serious consideration or transformation into something better. In later writings Wang went on to situate Wittgenstein's remarks against the backdrop of a wider, schematic distinction be-

---

p. 339). Moreover, "if a machine is to do mathematics, it is necessary that methods of logic be explicitly included" ([1961b], p. 332). This was an excellent forecast when it was written.

tween practice versus theory ("doing and being") in the philosophy of mathematics. He was thus able to see Wittgenstein early on as a critic, rather than an advocate, of "linguistic" philosophy, one who took seriously the realm of the familiar, but nevertheless made useful reflections on meaning and language, some of which could be turned in a constructive, practical direction in, e.g., computer science. This, Wang held, shed critical light on the limitations of reductionist ways of approaching "foundations" of mathematics. Of course, taken as a comprehensive view, Wittgenstein's would be unnecessarily restrictive. But just this illustrated, for Wang, the importance of aiming for a kind of comprehensiveness that would be useful, but partial.

Wang's "paradox," called by Wang himself "the paradox of small numbers" and an instance of the sorites paradox, was made famous by Dummett [1975]. The conceptual difficulty is alluded to, though not explicitly stated, in Bernays [1935], p. 60, and in Wang [1958a], p. 473 (which cites Bernays): there is no precise limit to be drawn between "feasible" and "non-feasible" (or "accessible" and "non-accessible") numbers, even though intuitionists proceed without regard to this limitation of idealization (or application of the induction schema). It may be stated as a paradox thus. Some numbers are, intuitively, small, and others large. If a natural number n is small, then so is $n + 1$. But then all natural numbers n are small, by induction.

The use of the induction schema is not viewed as essential to the paradox by Dummett [1975], who also shows how to formulate it with observational predicates such as "red" and "apodictic"—arguably concepts we do not apply *via* inference. Dummett argues that the paradox shows that strict finitists are committed to an inconsistent pair of beliefs, and so, more generally, are any users of a language whose vague concepts display a similar absence of a clear boundary in application. When generalized to treat observational terms, or the notion of "perspicuous," as in Wang ([1991?], p. 86), it seems that there is a systematic problem given the demand for consistency: concepts that throw up sorites-type phenomena are incoherent (cf. Wright [1980], p. 137).

Strict finitists, for example, hold that the meanings of all terms in mathematics must be given in relation to constructions that we are capable of actually effecting, and depend upon our capacity to recognize such constructions as providing proofs of those statements. But they must then also hold that there are non-empty sets of natural numbers (of, e.g., those that are "small" or "apodictic") which are closed under the successor operation, but also are bounded above.

The paradox did not pertain for Wang directly to the theory of meaning, as it does for Dummett, but instead to the theory of knowledge. This, however, as we have seen, is in Wang's hands a complicated matter. His position is not to be equated with contemporary epistemicist interpretations of the paradox, which hold that there is a point where the successor step breaks down, but we do not (yet) know where it is. For Wang the paradox shows that our responses to the use of a vague predicate are to be tested against two fundamental parameters, the "formal" and the "intuitive," and reveal the one-sidedness of a demand for global consistency proofs. Our experience and our concepts cannot be fully reduced to the formal.[30] And yet this does not mean that a formal rigorization of the notion of a "feasible" number might not be of interest, as in the first-order theory of bounded arithmetic (cf. Parikh [1971], Buss [1999]).

The paradox was perhaps of more importance to Wang than Dummett suggested at the end of his [1975], where he acknowledged Wang's framing of it. For Wang the paradox shows the inherent vagueness surrounding our uses of the notions of "perspicuousness," "smallness of number," and "understanding" (cf. [1991?], p. 86). But it thereby shows, not the need for a theory of the meaning of vague predicates, but instead the force of a kind of Platonism about the totality of numbers and the difficulty of attempting to

---

[30]In this respect Wang's attitude bears comparison to the later account of Wright [2007]. But Wang did not reject the possibility of articulating a rule that might be followed in applying vague or other predicates. Nor did he propose a special logic of negation (intuitionistic, for example) to handle the difficulty (cf. Putnam [1983], Dummett [2007]). He chose instead to reject the demand for a univocal or reductive application of a formal theory of deduction to a series of perceptual (or "intuitive" mathematical) experiences. As an historical point, Wright's prefatory remarks (p. 415) about philosophers in the 1950s generally viewing vagueness as "a marginal, slightly irritating phenomenon," attended to only by those who pursued "the amateur linguistics enjoyed by philosophers in Oxford in the 1950s" and idealized away by those favoring the use of formalized languages, do not apply to Wang.

170

adhere to what is "really" intuitive given the vagaries of language (cf. LJ, pp. 212-213). It also reinforced his idea of the purpose- and interest-relativity of formalization. Thus, in general, it illuminates the importance of the concrete and the complexities of its relation to theorizing. It allows the notion of meaning to be approached with subtlety, and the distinction between confusion and contradiction to be sharpened in its applications.

Mathematical induction is a schema, a "formal" method used in proofs. It may produce "paradoxes" when applied directly to natural language, but the contradictions that arise here should not engender skepticism. They should rather lead us to reflect on the limits of idealization and the need for it. The paradox suggests that there is "a gray area between the analytic and the synthetic" (BAP, p. 128), according to Wang. The idea that we could single out the operation of adding 1 as a clear basis for getting "larger" and "larger" numbers is itself "an act of abstraction, to give form to a range of nebulous relations of order" (LJ, p. 213). The use of the paradox is to illuminate that this is a specific step, and one that is not self-evident. It is also to unmask the misplaced nature of demands for proofs of absolute perspicuousness and/or consistency in the general foundations of knowledge or meaning. FMP attributes to Gödel an appreciation of this point (p. 44).

4. Factualism and Conceptualism

By "facts" Wang intended something "gross," not something meta-physical (FMP, p. 3). Wang's "factualism" contains more than a touch of Moore, at least in the sense that Wang believed that philosophy should not be allowed to override all that is familiarly taken to encompass "what we know," and that reminders of what we know serve a crucial philosophical function. Yet unlike Moore, Wang stressed how difficult it is to do justice to what we do know, and he embraced a general fallibilism, even while wanting to hold that we possess at least some knowledge that may be regarded as unconditional. He also insisted that facts may be meaningfully understood and discussed only from within wider structures of their interconnections, which he was willing to call, in a very broad sense, "logical." His philosophy contains nothing of the inveterately plain man or the particularist; Wang was a "conceptualist." Like Dedekind, he believed that our knowledge of number, as autonomous, cannot possibly be seen to stem from any particular sensory experience or capacity for empirical knowledge, but springs from and concerns concepts or intensions alone, and these insofar as they figure in law-governed structures of pure thought, rather than knowledge of the physical world or the structure of space as we perceive it.[31] Of course, Wang appreciated the point that "law-governedness" is a contentious notion in philosophy, where it receives a variety of interpretations. His interest in Wittgenstein partly turned on his efforts to grapple with this contentiousness, though it is not clear exactly how he resolved it in the end.

What is clear is that Wang took this broad form of Dedekindean anti-empiricism and revision of Kant's philosophy of mathematics as a starting point. His reasons for resisting empiricism thus lay in his knowledge of modern mathematics, and the seriousness with which he took the challenge of incorporating into proof reasoning about infinitary objects. Like Dedekind, Frege, and Gödel, Wang took the laws of number to arise, if not autonomously, then independently of any particular sensory experience or form of possible human sensation.

Wittgenstein would have agreed with this conclusion, but he argued for it very differently, and the contrasts with Wang's methods of argumentation are instructive. First, Wittgenstein questioned the idea that Dedekind's construction of number forces the philosopher into any particular position on the nature of mathematical objects generally. Second, he argued by elementary example, devising numerous pictures and images which fail to secure on their own a significant mathematical application or proof (cf. [1956], Part I). His general point is to indicate that no mental account of representations, perceptions, or impressions is able to account for our knowledge here—not, as Wang held, that appeals to inner mental life are, ideally, to be banished from philosophy altogether (cf. [1991?], pp. 170ff.; LJ, p. 356). Wittgenstein's point involves neither reductive behaviorism nor reductive appeal to sociological facts, but instead a philosophical generalization of Frege's context principle: a picture or image or impression only has meaning within an artic-

---

[31]Wang [1957a] is still a model of pedagogical clarity about Dedekind's contributions to the theory of number in their historical context.

ulated proof or systematic structure of articulated procedures.

Yet Wang, suspicious of appeals to language, was inclined to find this generalization of a Fregean approach inadequate.[32] This was partly because he aimed to bring logic and philosophy into accord with psychology. In his latest work, Wang felt that "what is at stake" in characterizing the nature of logic "may be construed as a determination of the universal receptive scheme of the human mind, which is to capture the underlying intersection of the diverse schemes actually employed by human beings, which are presumed to be potentially convergent" ([1996a], p. 366). The propositions of logic, he wanted to agree with Wittgenstein (in a remark he quoted from *On Certainty* (§401)), "form the foundation of all operating with thoughts" (LJ, p. 367). But Wang meant this literally, as part of the foundations of psychology. Wittgenstein was by contrast more likely to regard assemblages of remarks about mental terms as part of an exploration of the effects of grammar and everyday speech on theory.

A focus on concepts from within a logical analysis of their role(s) was, however, attractive to Wang. And from within that focus, he was content to speak of "intuition," especially where he focused on the "perspicuous." Though he appreciated Wittgenstein's efforts to find "perspicuous," easily grasped renditions of fundamental philosophical problems, he always read him in a behavioristic vein, just as had Bernays [1959]. This created a kind of looming tension, if not inconsistency in his reading of Wittgenstein. Wang knew that Wittgenstein considered talk about "intuition" as misleading and likely to be harmful. But he felt that Wittgenstein relies everywhere on "introspection" and the appeal to (his own) "intuitions," and so cannot deny their presence ([1991?], p. 171).

Where in the end there was some apparent meeting of minds was in Wittgenstein's willingness to treat philosophy as an investigation of the very notion of "concept" itself, as well as apparently fundamental philosophical concepts. Wittgenstein's later uses of language games allowed for a flexible and fallibilistic approach to

---

[32]For this reason Wang rejected Tait's deflationary idea that Wittgenstein's attack on the Augustinian picture of reference justifies a conception of the Platonism of mathematics as a "truism" (cf. Tait [2005], chapter 3; Wang [1991?], p. 2). His Platonistic reading of Frege, indebted to conversations with Palle Yourgrau, shaped his responses to the issues in [1991?].

the fundamental notions that attracted Wang.

Wang was explicit that, in following Gödel's idea of "concept," he did not want to prejudge the issue whether there are concepts, or rather in what sense there are concepts ([1991?], p. 161). Attention to what we in fact do helped him make sense of the idea of such an approach. And perhaps it allowed him to proceed with conceptualism although the foundations of this position were not fully clarified. Wang never worked out what he regarded as a satisfactory account, either of the notion of "concept" or of "intension": he was well aware of the plurality of views expressed in this regard and shied away from any definite account. Here too he resembles Wittgenstein, who was an inveterate critic of extensionalism insofar as it is regarded as a sufficient or fully adequate point of view, while at the same time lacking any firm account of the notions of concept or meaning.

Wang certainly did not think there was any such thing as an "intuitive" grasp of the notion of concept itself, even if certain particular concepts, such as that of computability, could receive "more or less definitive" analysis by means of "intuitive" representations (cf. [1991?], p. 139).[33] He seemed to think that any play in the notions of intension and concept could be narrowed down by beginning with at least the broadest features of what is taken to be known. And he read Wittgenstein, with his notion of language game, as having agreed here. An interesting feature of his emphasis upon looking at mathematical practice is that it tempered his understanding of Gödel's view of concepts. On Wang's reading, Gödel's "Platonism" and conceptualism are not brute, but more holistic and fallibilist (cf. Shieh [2000]).

Wang's fallibilism and commitment to a broad, pluralistic conception of logic gave him at least the beginnings of a reply to the obvious objection, How are we (and who are we) to get at or determine what the facts *are*? As Wang saw it, each individual inevitably operates with his or her own range of immediate reactions,

---

[33]To justify this claim Wang quotes from Turing [1937], where Turing says (§9) that he makes a "direct appeal to intuition" (Wang [1991?], pp. 139 ff.; cf. FMP, pp. 81-5, 90-95). I do not think, however, that Turing had in mind any notion resembling Wang's, but a much more down-to-earth idea of "common sense," and in fact one indebted to Wittgenstein. On this see my [forthcoming a] and [forthcoming b].

always wearing subjectively tinged glasses, and in one way this is
the starting point of philosophy (LJ, p. 355). Rejecting Wittgenstein's distrust of the notion, Wang called these "intuitions." He
considered one of the most important aims of philosophy to be
the development of better immediate reactions in and for all of us.
And for this philosophical task to have sufficient material to begin,
Wang held that there must be a certain degree of universal com-
monality of structure to these intuitions. Modelling his account on
the *Grundlagenstreit*, Wang held that familiar facts of elementary
arithmetic ("2 + 2 = 4") form an important example here. Here
the influence of Bernays is felt: Bernays had also maintained that
knowledge of elementary arithmetic is of intuitive, or special evidential origin. But once again the philosophical frame Wang would
ultimately bring to bear on the idea was generalized. "Immediate
apprehension," or "intuition," could be "sensation, knowledge, or
even mystical rapport," according to Wang (LJ, p. 372).

5. The notion of "intuition"

Wang's notion of "intuition" is perhaps the greatest stumbling block
for readers trained in twentieth century Anglo-American philosophy, and it certainly marks the place where I myself find his philosophy most difficult to understand. He was well aware of Wittgenstein's and his likely readers' antipathy to the notion. Sometimes
he would insist that the notion of "intuition" could be eliminated
without loss from Gödel's or Wittgenstein's philosophies. Sometimes he used the notion in an everyday way, as a mathematician
does when looking for an "intuitive" way of seeing a proof (or parts
of a proof). Nevertheless, Wang's particular uses of the notion are
quasi-systematic and central to his own thought and to his accounts,
not only of Wittgenstein, but other philosophers, including Gödel.

The traditional European division of philosophy by way of a
distinction between "theoretical" and "practical" branches of the
subject was foreign to the explicit organization of faculty structures and specializations in Anglo-American philosophy of Wang's
day, but remained widespread in many other parts of the world
and in American curricula influenced by the German model. This
distinction was natural for Wang, although as he used it, it was be-

ing investigated, sharpened and extended, encompassing at times Eastern as well as Western philosophy. It did not correspond, for example, to a distinction between ethics and political philosophy versus metaphysics or philosophy of science. Instead, Wang often articulated its significance in terms of a "dialectic" or continuum or pair of forces, rather than an overarching dichotomy. He saw this dialectic as a driving force in the history of science given the importance of mathematization and idealization. But the distinction itself expressed a conviction that conceptualization always leaves something out. Wang's notion of "intuition" is not, however, a theory of non-conceptual content: the degree and extent to which "intuition" can become conceptualized is left open, as is the degree and extent to which concepts can be concretely presented in "intuition." Wang lacked a theory of perception, basing his philosophy on particular examples of our grasp of mathematical truth and on what Wang took to be everyday experiences of meaning, knowledge, and life.

Wang read Wittgenstein as holding, in his early philosophy as well as later on, that something intuitive always remains after conceptualization. This is Wang's understanding of the distinction between showing and saying, and it shaped his understanding of Wittgenstein, earlier and later, from his very first writings. In his review [1945] of Russell's *Inquiry into Meaning and Truth* he was already suggesting that philosophers should seek "possible sources of knowledge besides language" ([1945], p. 147 of translation). And in his latest writing he takes the novelty of Wittgenstein's approach to have been that he "begins and ends with the perceptual immediacy of our intuition of the actual use of words in a given situation," an approach Wang saw as a "way of pursuing the traditional quest for certainty in philosophy" (LJ, p. 329).

The later Wang takes there to be a "familiar gap" between "seeing and saying," between the understanding of a thought and the clarification of its meaning, its grasp vs. its communication. Because of the "subjective and fluid character" of seeing, philosophers can be driven to assure communication by "a direct appeal to the connection between words and deeds, to bypass the interference from passing through the mental" (LJ, p. 356). This bypassing, he believed, is illustrated in Wittgenstein's later thought. As to philosophical method, Wang emphasized that "saying... is only one way of communicating":

> Literature, for instance, tries to show the universal by say-
> ing the particular; similes and metaphors show one thing
> by saying something else; action, tone, and gesture can be
> shown in a drama or film but they can only be said or told
> in a novel... (LJ, p. 356)

This mode of communication was, he rightly felt, of central importance to Wittgenstein—although it is questionable whether Wittgenstein required a theory of non-conceptualized content to forward it. In his latest writings on Wittgenstein and Gödel, the connection between aesthetic experiences, intuitive presentations, and philosophical method are stressed by Wang. In fact, the appeal of conceptual thinking as such is its promise of finding for us "a perspicuous view of a larger whole" ([1991?], p. 172). But Wang seemed to feel by the end that philosophy required literary formulations to achieve its full potential to communicate concepts.

Thus unlike Bernays, whose use of the notion of "intuition" had a more specifically Kantian cast, Wang's notion is tied, at least when connected to Wittgenstein, to something like "respect for the particularity and immediacy of everyday experience." In epistemic terms, Wang defined it as a "summary 'perspicuous' grasp of massive details" ([1991?], p. 124). In this way, partly inspired by Wittgenstein, he followed Bernays's and Gödel's idea that Kant had been wrong to draw a sharp distinction between concept and intuition, but he held on to the ambition, common to Husserl, Gödel, and Wittgenstein, to use philosophy to clarify experience of everyday life. Wang put forward here a view intended to contrast with forms of conventionalism about analyticity and meaning, such as Carnap's, that he despised.

In his later writings Wang emphasized that Wittgenstein's *On Certainty* had taken up the subtle challenge of beginning with the actually (not the possibly) familiar, bringing it into the orbit of our understanding of logic itself. Wittgenstein, he believed, had made substantial progress beyond Moore in sketching how general philosophy might articulate a very broad conception of the logical as both adjudicative and able to unearth the structure of what is known unconditionally. This went well beyond a doctrine of "common sense" that involves the risk of foot-stamping. Wang took Wittgenstein to have successfully refined, especially in *On Certainty*, our understanding the problem of where philosophy should begin, and where it should end. The answer, as Wang understood it, was in logic, conceived as a broadly adjudicative inquiry. But Wittgenstein's examples ("Here is one hand", "This is green") displayed a practical, concrete, empirical or "intuitive" element in knowledge, hence in logic itself. Factualism could be seen to incorporate this method within itself, without becoming dogmatic. Our capacity for clarity and knowledge of everyday facts could be a starting point, but be treated, on any given occasion, as fallible.[34]

Wang's use of the notion of "intuition" also gave him a way to use Wittgenstein to resist and delimit Husserl's approach. There are a variety of notions of rigor in philosophy, and not all of them are to be equated. Wang regarded his own notion of "intuition" as a useful counterweight to Husserl's phenomenology, whose impact on Gödel he did not see as uniformly salutary. Husserl, for Wang, hoped to build up objectivity from an immediately given, subjective, inner experience, but this Wang deemed wrongheaded and in fact "disagreeable," both because it failed to begin with what we know, intersubjectively, at the start, and because it encouraged, he believed, conceptions of infallible intuitive insight and experience (BAP, p. 37). Wang did not believe that philosophy could aspire to become a "rigorous science" in the sense of a system (FMP, p. x). Here he differed with readers of Husserl, such as Føllesdal, who commended Husserl for his focus on problems of perception and his fallibilist conception of intentionality (LJ, pp. 350-352). For Wang, empiricism was a non-starter in light of the fundamental importance of the philosophy of mathematics and there was no general theory of intentionality apart from an account of the nature of logic. Perception was an interesting phenomenon for psychologists, but not for philosophers. Here too we see Wang's affinities with Wittgenstein.

"Intuition" is an attempt to express whatever the formal, theo-

---

[34]The view might be seen to have affinities with those enunciated in McDowell [2011], in terms of identifying a place in the space of possible positions for the manifestation of perception as a capacity for knowledge. Neither McDowell, nor Sellars, whom McDowell is representing, relied on a wider frame of intuitive vs. conceptual knowledge, although Sellars' idea of "the manifest image" bears an interesting comparison to Wang's later ideas. Yet neither Sellars nor McDowell are, like Wang, attempting to integrate an account of our knowledge of infinitary objects within the orbit of their theories.

retical, conceptualized misses or fails to represent. Wang admired Wittgenstein's criticisms of Russell on the notion of "self-evidence" in logic, and took to his resistance to logicism regarded as a program to eliminate the use of intuition in arithmetic. For Wang, intuition is not determinative, and it carries no form of justification with it. There is no gap in its application to the world or the facts. But it is not always to be regarded as less reliable than proof, much less as merely psychological.

Still, "intuition" is intended to fit within a general account of human mental receptivity as such. And here there does seem to lurk a genuine difficulty with Wang's views. Sometimes a formal principle such as mathematical induction is exactly what gives us an "intuitive" perspective on a property, and it is the particular cases that seem, collectively, obscure. Sometimes what seems transparently grasped through purely formal means is not—as when we see that the principle of mathematical induction is, perhaps surprisingly, actually logically equivalent to the pigeonhole principle. It is hardly the case that the formal principle never lies on the side of the practical, the intuitive, the concrete, or the particular. The boundaries of Wang's dichotomies are permeable, and in fact liable to shift: they are useful, but occasion sensitive. Thus the "dialectic" of the formal and the intuitive is not explanatory but parasitic on particular cases. And this has methodological importance for the philosopher. Wang held that "felt psychological certainty should be appealed to in a non-parochial way" ([1991?], pp. 79-80). But one might better hold, with Wittgenstein, that our handle on psychological concepts and semantical theories, even on the notion of proof in mathematics, requires the parochial (cf. Travis [2006]).

Yet the notion of "intuition" has a broad meaning in Wang's hands. It is not to be identified with anything more, or anything less, than the immediate, considered *appearance* of truth (to us), or certainty, rather than truth itself. It thus serves as an entering wedge for the beginnings of epistemology. As in Wang's readings of Gödel, "intuition" is, though immediate, fallible, liable to revision in light of reflection, and attaches to judgments, rather than to objects as such. In a sense, Wang's uses of the notion of "intuition" stick closely to the Kantian framework, within which the notion is tied, both to immediacy of representation and to the instantiation or exemplification of a concrete particular.

The purpose of philosophy is, however, tied directly to the notion for Wang: philosophy is to explore and to structure what is shared despite differences in our "intuitions," to make immediate, "perspicuous" sense of the rich but complex experiences of life. To achieve this one must also use and devise "intuitions," perhaps even aesthetic presentations, to communicate concepts. Mathematics, especially the familiar areas of elementary arithmetic on which Wittgenstein focused, offer a starting point from which the "factualist" may begin by reflecting on the nature and character of at least nearly universal agreement, and to titrate and measure differences as they appear. Positions in the philosophy of mathematics do not of course exhaust mathematical knowledge. But philosophy, even philosophy of mathematics, is not complete without providing something "intuitive," concrete, and clear that is extra-mathematical.

Wang thus had no use for direct forms of Platonism which picture us intuiting objects of knowledge directly with the mind's eye. Here he was certainly sympathetic to Wittgenstein's doubts. His use of the notion of "intuition" was not accompanied by any theory of perception. Yet there are tensions in his thought in this regard. Wang tended to think of Wittgenstein's notion of aspect perception as a matter for the psychology of vision rather than a theoretically basic phenomenon that could illuminate the notions of "intuition," "understanding," "concept," or "perspicuousness." Thus his reading of Gödel reflects a broadly logicizing view of "intuition"—but only so long as one understands that Wang's notion of the "logical" covers more than formal or symbolic logic, encompassing a variety of methods of analysis and elucidation (cf. Shieh [2000]).

In Wang's philosophy there is a constant adverting to the uncomfortable idea of a mystical or unutterable content that eludes conceptualization altogether. While theorists of perception who advocate the idea of non-conceptual content tend to rely on psychological data, avoiding commitment to such a paradoxical treatment of experience, it is not clear that Wang could do so, insofar as he was offering an account of philosophical experience itself. To him this risk of mysticism was all to the good, and connected his philosophizing with certain strands of Buddhism, as well as Gödel's thought. But it seems clear that at least Wittgenstein, by contrast, had far greater faith in the effort to use our means of representa-

180

tion to accurately reflect on, and philosophize about, the structure of our thoughts.

It is interesting that since Wang's death there has been a revival of interest in the structure of appeals to "intuition" in analytic philosophy, as conceptualism and the development of general metaphysics have been revitalized. Non-conceptual content is discussed as part of philosophical method much more frequently than during Wang's lifetime. Within epistemology, the relation of "knowing how" to "knowing that" receives much more attention than heretofore. Perhaps on these scores Wang was ahead of his time. It would be interesting to analyze in another place how his appeals to intuition compare with those of recent theorists, and to see how much farther they have gotten than he. This would be the real way to assess the worth of Wang's philosophy.

6. Philosophy as a Quest for Perspicuous Objectivity

Throughout Wang's philosophy, as we have seen, the most central philosophical problem is what might be termed "the problem of disagreement," the (partly practical) problem of how philosophy might go about distilling or "decomposing" a range of possibly universally held, agreed upon judgments from the fact of multiple disagreements between individuals with different immediate reactions, feelings, histories, experiences, and "intuitive" perceptions. He was less interested in what might be termed "the problem of agreement," which also bothered Wittgenstein, namely, What if the agreements we express cover up profound misunderstandings?[35]

The epilogue to Wang's LJ (Chapter 10) assesses what progress he thinks was made in twentieth century philosophy on this problem and assesses prospects for the future. It is unsurprising that Wittgenstein, whose notion of "agreement in judgment" is notoriously central but problematic, would play a central role in Wang's thoughts here. For Wang, the later Wittgenstein saw the "natural"

---

[35]If one sees this as a problem for Wittgenstein, then Wang's picture of him as a conventionalist or ordinarily language contractualist about the notion of "agreement" is unattractive. See Wittgenstein [1980], §1107, for an explicit place where it is asked, of a perceptual case, "What if this full agreement ("It's like that for me too!") were based on a misunderstanding?" See also [1953], §241, also pp. 226-27 and 230.

inclination to generalize as the main source of confusion in philosophy, and Wittgenstein's recommendation, a novel one, was to insist that philosophy begin and end "with the perceptual immediacy of our intuition of the actual use of words in a given situation" (LJ, p. 329). This methodological way of understanding Wittgenstein's notion of the "perspicuous" came more and more to the fore in Wang's later writings, as it was taken by him to show the importance of concreteness and practice for knowledge.

In fact in the unpublished manuscript [1991?] Wang defines philosophy as "the quest for (comprehensive) perspicuous objectivity" (p. v). By focusing on a "dissection" of conceptual objectivism in mathematics, he hoped to point toward a "flexible frame" for philosophy. And his initial idea was that the "less audacious parts" of the philosophies of Gödel and Wittgenstein might be excerpted, seen to be compatible, and used to erect that frame. In LJ there is much overlap with this manuscript but an important difference, in that in this later, more mature work, the comparison between Gödel and Wittgenstein is absorbed into a wider book framework, reduced to an explicit chapter and then folded into more methodological remarks in a final epilogue.

Wang believed that philosophy progresses most fruitfully by clarifying, presenting, and resolving differences intuitively and perspicuously, though not by restricting itself to the actual uses of words. The aim should be to alter the interplay between practice and theory while not promising forms of certainty that cannot be reasonably achieved. Philosophy should then ideally begin with what is familiar and agreed upon, and move from there, cataloging, analyzing, and decomposing disagreements as they emerge. And it should be held, at least eventually, to the aim of altering everyday life, as well as theory.

7. Wang's style and his final motto

In addition to the subtlety of overlaying Wang's aims upon Wittgenstein's, there is the challenge of penetrating Wang's own style, which, overlaid on Wittgenstein's—itself hardly a usual or clear one—is not easy to characterize.[36] Wang's philosophical goal was

---

[36]Here I agree with Parsons [1998] and Shieh [2000].

to draw a portrait of a philosopher or view and then to identify which of its features he found more, and which less, attractive, thereby revealing his self, freely admitting from the start that his own immediate "intuitions" and philosophizing shaped his reactions. Wittgenstein's later method was to react to his own (and others') remarks in a multilogue, polyphonic interlocutory fashion, carefully orchestrated for literary effect, in order to represent the nature and character of philosophical thought. When arguments are found in Wang, they are deductions from his observations, or relative comparisons of the interest and correctness of different responses, concepts, or principles. The tone of the whole is always tentative, pluralistic, and, as Parsons [1998] has emphasized, synoptic rather than systematic in aim.

Yet Wang's literary ambitions became stronger over time, and importantly shape his final writings. These bear an important relation to his ideas about "intuition" and his interest in Wittgenstein, whom he came to regard as "art centered" rather than "science centered" in his conception of philosophy (BAP, p. 75). Wang felt the literary effects of Wittgenstein's writing were not irrelevant to the content of his philosophy. I think we can assume that Wang felt the same way about his own books.

In the manner of Walter Benjamin,[37] or perhaps better, of his Chinese forebears, Wang often proceeded by arranging quotations, without interpretation, in an effort to draw out the reader's reflection and response, thereby showing, but not himself directly stating. Wang's attachment to this way of writing philosophy indicates one reason he may have found Wittgenstein's later interlocutory style less off-putting than many philosophers do. He liked its intuitiveness, and he modeled his own writing on its way of inviting reflection without necessitating dogmatic adherence. It showed rather than stated, was intuitive rather than theoretical. Wang himself sought to probe philosophical systems dialectically and intuitively, without an overarching theory. He was explicit that he considered his own approach to be "more effective" than Dummett's in the quest for clarification of differences and distinctions of meaning ([1991?], p. 131).

The device of quotation, used by Wittgenstein at the opening

---

[37]This is recounted by Arendt in her introduction to Benjamin [1969], p. 4.

of *Philosophical Investigations*, was one Wang admired and often used, both when discussing Wittgenstein and when attempting to characterize his own views. His mottos are often useful clues to his thoughts.

As early as [1958a], when introducing anthropologism, Wang quotes from Butler the saying Wittgenstein once considered taking as a motto to the *Philosophical Investigations*: "Everything is what it is, and not another thing," thus endorsing the importance of attending to particulars and their differences from one another.[38]

He closes chapter 2 of BAP with an interpretation of the end of the *Tractatus* and Wittgenstein's form of "mysticism" by drawing a comparison between the early Wittgenstein and the Buddhist Wu Yunzeng, both of whom are, for Wang, "more concrete" than traditional philosophers like Plato, Aristotle, conceptualizers of Tao, and Spinoza when it comes to thinking about traditional ideals of an unattainable limit (BAP, p. 100). He does this by means of a quote from Wu Yunzeng.

The structure of quotations and mottos in LJ is perhaps most illuminating of his style. In Wang's opening motto to this work, he pairs a quote from Hegel about objective logic embracing the "wealth of the particular" with a quote from Lu Jiuyuan stating that the minds of all sages from the East and the West "have the same kind of intuition." The introduction that follows this allusion to universalistic particularism opens with another contrasting and interlocking pair of quotations, this time from Gödel and Wittgenstein, concerning generalization in philosophy. Gödel's begins: "Philosophers should have the audacity to generalize things without any inhibition: go on along the direction on the lower level, and generalize along different directions in a uniquely determined manner." Wittgenstein's runs: "Hegel seems to me to be always wanting to say that things which look different are really the same. Whereas my interest is in showing that things which look the same are really different."

Wang's epilogue, or final chapter, opens with four quotations concerning knowledge and understanding, designed, it seems, to show the points where his reflection was directed at the end of his

---

[38]As Wang knew, Wittgenstein had considered for the motto of the *Investigations* the saying from Lear, "I'll teach you differences!" (Monk [1990], pp. 536-37).

life. One is from Confucius's *Analects*, one from Bernays's essay on rationality [1974], one from Wang's own *Beyond Analytic Philosophy* ([1985a], quoted above as my motto for this essay), and the last is from Rawls's *Political Liberalism* [1993], a remark about the method of "reflective equilibrium."

Wang's final motto for his own corpus, taken as my own motto in this essay, is obviously intended as a critical comment on Kant's characterization of the questions that had unified his own critical project, namely "What can I know? What ought I do? For what may I hope?"[39] To orient us with respect to Wang's overarching philosophical aims at the end of his life, let us compare and contrast the structures of these two quotations.

The first and perhaps most important thing to note is that Wang offers a "classification of what philosophy has to attend to." The suggestion is that philosophy will have to go outside of itself in order truly to know itself: the subject is not autonomous, or self-authenticating.[40] Wang takes there to be extra-philosophical matters that no philosopher can or should ignore or repress. So the central focus and concern of philosophy is not itself purely philosophical, as, at least arguably, was Kant's. Wang does not offer a list of questions bound to generate ultimately unsolveable conundrums and a focus on a transcendent, "higher" world of hopes and dreams. He also does not claim that his classification is exhaustive of any interests of reason or philosophy, as does Kant: he frames a "guiding principle" or rule of thumb, rather than an *a priori* characterization of superordinate challenges or questions. Finally, Wang's use of the idea of "doing justice" is intentionally amorphous and flexible, yet central to his own conception of philosophy, which in his own mind bore, even if indirectly, on ethical, moral, and political questions. His philosophy aimed to cope, practically and intellectually, with the fact that there are incompatible perspectives, as well as the facts of reality as science knows them.

---

[39]"All the interests of my reason, speculative as well as practical, combine in the three following questions: 1. What can I know? 2. What ought I to do? 3. For what may I hope?" (*Critique of Pure Reason* A804-05/B 832-33.)

[40]As Wang elsewhere wrote, to know mathematics we must go outside mathematics to the world, by way of our experience. Thus philosophy of mathematics is not self-sufficient as a philosophy. Ending the manuscript in which he hoped to compare Wittgenstein and Gödel, he quoted Gödel: "in order to know what mathematics is, one has to know what the world is" ([1991?], p. 175).

Like Kant's characterization of his philosophy, Wang's is tripartite. The first element of his guiding principle concerns knowledge. His "substantial factualism" and "conceptualism" aimed less at a limiting of knowledge, or discussion of possible knowledge, than at the faithful and clear exposition of what we *do* know. This shifts the modality of Kant's question in a way clearly appropriate to the contemporary world in which the quantity of information and knowledge increases faster than the ability to survey it and lay down *a priori*, universal principles for its justification. It also states a constraint on philosophizing that Wang felt Wittgenstein had come too close to violating: never restrict what we know in light of a philosophical theory. Of course, this constraint also resonates with a part of Wittgenstein Wang came to emphasize, especially in his later writings: the importance and difficulty of beginning with what is familiar, shared, and actually given to us.

What of the second element of Wang's "guiding principle"? Here Wang suggests that philosophers replace Kant's call for specification of ethical action with an exploration of "what we believe." The problem of what in fact we do believe, the task of openly specifying where and how we disagree and agree, was central to Wang's conception of philosophy, in practice and in theory. He was inclined to think that twentieth century history, including both politics and the history of science, had shown the importance of open exchange and discussion as prior to action, as well as the limits of philosophy treated as a purely foundational enterprise. Finally and thirdly, there is Wang's interest in "how we feel." This echoes Kant's idea in his (third) *Critique of Judgment* that the bridge between theory and practice in the critical philosophy may built through attention to aesthetics, emotion, and art. While Wang's writings about psychology and aesthetics are neither extensive nor especially penetrating, he did take the subjects seriously, and was interested in the "queer resemblance" Wittgenstein once claimed to have perceived between an investigation in mathematics and one in aesthetics.[41]

---

[41]In fact, Wang ends his [1991?] with a digression on the "queer resemblance" Wittgenstein noted between an investigation in philosophy ("perhaps especially in mathematics") and in aesthetics (Wittgenstein [1998], p. 29e; MS 116,56 in [2003]), and a comparison with Gödel's views. In the early 1990s I frequently spent time in the enjoyable company of Wang and his wife Hanne Tierney, an accomplished artist (http://www.hannetierney.com/artist-bio.html).

Wang read persistently in psychology, viewing it as an open and developing science to which philosophers should attend; he would surely have had much to say about the recent rise of "experimental philosophy," and probably not have wanted it dismissed out of hand. In general Wang had no use for an eliminative view of psychology in philosophy, including philosophy of mathematics. This is evidenced by his occasional remarks about Freud's importance and his knowledge of, and praise for, books that concern the psychology of invention and discovery in mathematics. His interpretation of Wittgenstein, insofar as it took to be fundamental the idea of aspect perception as part of the notion of *perspicuousness*, turned on a notion that in the eyes of a Fregean, or for that matter a contemporary reader of Wittgenstein on rule-following, is suspiciously phenomenological, a potentially subjective or quasi-psychologistic admixture out of place in discussions of logic and mathematics. For Wang, it was admirable that Wittgenstein had toyed with the idea of taking his later writings to offer an investigation of the foundations of psychology, and important that he had been knowledgeable of at least some psychological results and experiments.

In general Wang took appearances, feelings, lived experiences, emotions, strikings, and so on as fundamental data in philosophy to which we must try to do justice. Against Wittgenstein, he called these "intuitions." Wang had a strong tendency to biographize, especially in his later work, but it must be said that the idea of setting a particular, exemplary philosophical life before the reader, familiar now in certain branches of virtue ethics, was part of what he felt a responsible historian of philosophy should do. Unlike most analytic philosophers, for Wang one test of knowledge is its usefulness for life.

8. Final thoughts

An historical note is in place here. I was a bystander to Wang's interest in Wittgenstein during the period 1990-1994, when I had the great pleasure of spending once a week, sometimes more, in philosophical conversation with him at Rockefeller University. He approached me with the idea of discussions just after my arrival

---

Wang was inclined to consider artists with as much respect as he did scientists.

as a young assistant professor at C.U.N.Y., asking for comments on section 6 of his [1991b], and knowing that I was writing on Wittgenstein's remarks on the foundations of mathematics. During this period he was planning a book, variously titled at various times, in which the comparison of Gödel and Wittgenstein would take center stage.[42] He was interested in learning more about the later Wittgenstein, and in discussing the present state of philosophy.

What was remarkable, as I reflect back on our conversations, was the tenacity, energy, supportiveness, and openness of Wang as a teacher. He was a good teacher by being a good listener. He had specific scholarly questions and his own body of writing he was working on, and he was interested in hearing what someone much younger than himself would say about his ideas and their own.

I am personally and intellectually greatly indebted to Wang's example, as a teacher, a colleague, and a philosopher. My own work on Wittgenstein has been deeply shaped by his, and his encouragement for that work, undertaken while I was a junior faculty member, was unstinting and generous. No young philosopher could have had a finer mentor. Hao Wang provides an authentic example to me of what the Chinese call *xiansheng*, and the Japanese *sensei*: someone who *knows* something.[43]

---

[42]The titles he discussed with me included *For Perspicuous Objectivity: Discussions with Gödel and Wittgenstein* (a manuscript with this title, cited here as [1991?], was given to me by Charles Parsons), and also *Gödel, Wittgenstein and Purity of Mind: Logic as the Heart of Philosophy*. Wang had great difficulty producing this manuscript, and in the end set the comparative project aside, cutting out some of the parts on Gödel and Wittgenstein later published in LJ. My sense is that he would have wanted to return to the comparative manuscript at a later stage, but he did not manage to do so.

[43][*Sensei*, like *xiansheng*, means Teacher or Master. Eds.]

References

Cited works of Hao Wang:[44]

1945. Language and metaphysics (Chinese). *Zhe xue ping lun. Philosophical Review* 10, no. 1, 35-38. Published in 1946. English translation, [2005].

1950c. On scepticism about induction. *Philosophy of Science* 17, 333-335. Reprinted in [1974a], Appendix.

1955e. On formalization. *Mind* 64, 226-238.

1957a. The axiomatization of arithmetic. *The Journal of Symbolic Logic* 22, 145-158.

1958a. Eighty years of foundational studies. *Dialectica* 12, 466-497.

1960b. Proving theorems by pattern recognition, part I. *Proceedings of the Association for Computing Machinery* 3, 220-234.

1961b. Process and existence in mathematics. In Yehoshua Bar-Hillel, E. I. J. Poznanski, M. O. Rabin, and Abraham Robinson (eds.), *Essays on the Foundations of Mathematics, Dedicated to Prof. A. A. Fraenkel on His 70th Anniversary,* pp. 328-351. Jerusalem: Magnes Press, The Hebrew University of Jerusalem.

1974a. *From Mathematics to Philosophy.* London: Routledge & Kegan Paul. Cited as FMP.

1984a. The formal and the intuitive in the biological sciences. *Perspectives in Biology and Medicine* 27, 525-542.

1985a. *Beyond Analytic Philosophy. Doing Justice to What We Know.* Cambridge, Mass.: MIT Press. Cited as BAP.

1987b. Gödel and Wittgenstein. In Paul Weingartner and Gerhard Schurz (eds.), *Logic, Philosophy of Science and Epistemology,* pp. 83-90. Proceedings of the 11th International Wittgenstein Symposium, Kirchberg am Wechsel, Austria, 4-13 August, 1986. Vienna: Verlag Hölder-Pichler-Tempsky.

---

[44]Information about reprintings is largely omitted here, but it is given in the full bibliography at the end of this volume.

1991b. To and from philosophy—Discussions with Gödel and Wittgenstein. *Synthese* 88, 229-277.

1991?. *Gödel, Wittgenstein and Purity of Mind: Logic as the Heart of Philosophy.* Unpublished manuscript.

1996a. *A Logical Journey. From Gödel to Philosophy.* Cambridge, Mass.: MIT Press. Cited as LJ.

2005. Language and metaphysics. Translation by Richard Jandovitz and Montgomery Link of [1945]. *Journal of Chinese Philosophy* 32, 139-147.

Other works cited:

Auxier, Randall E., and Lewis Edwin Hahn, 2007. *The Philosophy of Michael Dummett.* The Library of Living Philosophers 31. Chicago and La Salle, Ill.: Open Court.

Benacerraf, Paul, and Hilary Putnam (eds.), 1964. *Philosophy of Mathematics: Selected Readings.* Englewood Cliffs, N. J. Prentice-Hall. 2d ed., Cambridge University Press, 1983.

Benjamin, Walter, 1969. *Illuminations.* Edited with an introduction by Hannah Arendt. Translated by Harry Zohn. New York: Schocken Books.

Bernays, Paul, 1935. Sur le platonisme dans les mathématiques. *L'enseignement mathématique* 34, 52-69. English translation in Benacerraf and Putnam [1964]. Both reprinted (with revision of the translation) in Bernays [forthcoming].

Bernays, Paul, 1959. Betrachtungen zu Ludwig Wittgensteins *Bemerkungen über die Grundlagen der Mathematik. Ratio* 3, 1, 1-18. English edition of *Ratio*, 2, 1, 1-22, reprinted in Benacerraf and Putnam [1964] (1st ed. only). Both reprinted in Bernays [forthcoming].

Bernays, Paul, 1974. Concerning rationality. In Paul Arthur Schilpp (ed.), *The Philosophy of Karl Popper*, pp. 597-605. The Library of Living Philosophers 14. La Salle, Ill.: Open Court.

Bernays, Paul, forthcoming. *Essays on the Philosophy of Mathema-*

190

tics. Wilfried Sieg, W. W. Tait, Steve Awodey, and Dirk Schlimm, eds. Chicago and La Salle, Ill.: Open Court.

Buss, Samuel R., 1999. Bounded arithmetic, proof complexity, and two papers of Parikh. *Annals of Pure and Applied Logic* 96, 43-55.

Dummett, Michael, 1975. Wang's paradox. *Synthese* 30, 301-324. Reprinted in *Truth and Other Enigmas.* London: Duckworth, 1978.

Dummett, Michael, 2007. Reply to Crispin Wright. In Auxier and Hahn [2007], pp. 445-454.

Floyd, Juliet, 1995. On saying what you really want to say: Wittgenstein, Gödel and the trisection of the angle. In Jaakko Hintikka (ed.), *From Dedekind to Gödel: The Foundations of Mathematics in the Early Twentieth Century,* pp. 373-426. Dordrecht: Kluwer.

Floyd, Juliet, 2000. Wittgenstein, mathematics, philosophy. In Alice Crary and Rupert Read (eds.), *The New Wittgenstein,* pp. 232-261. London and New York: Routledge.

Floyd, Juliet, 2001. Number and ascriptions of number in Wittgenstein's *Tractatus Logico-Philosophicus.* In Juliet Floyd and Sanford Shieh (eds.), *Future Pasts: The Analytic Tradition in Twentieth-Century Philosophy,* pp. 145-191. New York and Oxford: Oxford University Press.

Floyd, Juliet, forthcoming a. Wittgenstein's diagonal argument. A variation on Cantor and Turing. In Peter Dybjer, Sten Lindström, Erik Palmgren and Göran Sundholm (eds.), *Epistemology versus Ontology: Essays on the Foundations of Mathematics in Honour of Per Martin-Löf.* Dordrecht; Springer.

Floyd, Juliet, forthcoming b. Turing, Wittgenstein, and types: Philosophical aspects of Turing's "The reform of mathematical notation and phraseology" (1944-45). In S. Barry Cooper and J. van Leuven (eds.), *Alan Turing: His Work and Impact.* Amsterdam: Elsevier.

Gellner, Ernest, 1959. *Words and Things. A Critical Account of Linguistic Philosophy and a Study in Ideology.* London: Gollancz.

Kreisel, G., 1958. Review of Wittgenstein, *Remarks on the Foundations of Mathematics. British Journal for the Philosophy of Science* 9, 135-158.

Marion, Mathieu, 2009. Radical anti-realism, Wittgenstein, and the length of proofs. *Synthese* 171, 419-432.

Marion, Mathieu, forthcoming. Wittgenstein on the surveyability of proofs. In Marie McGinn (ed.), *The Oxford Handbook to Wittgenstein*. New York and Oxford: Oxford University Press.

McDowell, John, 2011. *Perception as a Capacity for Knowledge*. The Aquinas Lecture, 2011. Milwaukee: Marquette University Press.

McGuinness, Brian, 1988. *Wittgenstein: A Life. Young Ludwig, 1889-1921*. Berkeley and Los Angeles: University of California Press.

Monk, Ray, 1990. *Wittgenstein. The Duty of Genius*. London: Jonathan Cape. Reprinted by Penguin Books.

Mühlhölzer, Felix, 2006. 'A mathematical proof must be surveyable': What Wittgenstein meant by this and what it implies. *Grazer Philosophische Studien* 71, 57-86.

Mühlhölzer, Felix, 2010. *Braucht die Mathematik eine Grundlegung? Ein Kommentar des Teils III von Wittgensteins Bemerkungen über die Grundlagen der Mathematik*. Frankfurt am Main: Vittorio Klostermann.

Parikh, Rohit, 1971. Existence and feasibility in arithmetic. *The Journal of Symbolic Logic* 36, 494-508.

Parsons, Charles, 1996. In memoriam: Hao Wang, 1921-1995. *The Bulletin of Symbolic Logic* 2, 108-111.

Parsons, Charles, 1998. Hao Wang as philosopher and as interpreter of Gödel. *Philosophia Mathematica* (3) 6, 3-24.

Putnam, Hilary, 1983. Vagueness and alternative logic. *Erkenntnis* 19, 297-314. Reprinted in *Realism and Reason: Philosophical Papers, volume 3* (Cambridge University Press, 1983).

Rawls, John, 1971. *A Theory of Justice*. Cambridge, Mass.: The Belknap Press of Harvard University Press.

Rawls, John, 1993. *Political Liberalism*. New York: Columbia University Press.

Shieh, Sanford, 2000. Review of Wang [1996a]. *Erkenntnis* 52, 109-

115.

Tait, William, 2005. *The Provenance of Pure Reason. Essays on the Philosophy of Mathematics and its History.* New York and Oxford: Oxford University Press.

Travis, Charles, 2006. *Thought's Footing.* New York and Oxford: Oxford University Press.

Turing, A. M., 1937. On computable numbers, with application to the *Entscheidungsproblem. Proceedings of the London Mathematical Society* (2) 42, 230-265.

Von Wright, G. H., 1994. Logic and philosophy in the twentieth century. In Dag Prawitz, Brian Skyrms, and Dag Westerstahl (eds.), *Logic, Methodology, and Philosophy of Science IX*, pp. 9-25. Amsterdam: Elsevier.

Wittgenstein, Ludwig, 1953. *Philosophical Investigations.* Edited by G. E. M. Anscombe, Rush Rhees, and G. H. von Wright. With a translation by G. E. M. Anscombe. Oxford: Blackwell.

Wittgenstein, Ludwig, 1956. *Remarks on the Foundations of Mathematics.* Edited by G. E. M. Anscombe, Rush Rhees, and G. H. von Wright. With a translation by G. E. M. Anscombe. Oxford: Blackwell.

Wittgenstein, Ludwig, 1974. *On Certainty, Über Gewissheit.* G. E. M. Anscombe and G. H. von Wright, eds. Translated by Denis Paul and G. E. M. Anscombe. Oxford: Blackwell. (Corrected edition; first published 1969.)

Wittgenstein, Ludwig, 1980. *Wittgenstein's Lectures, Cambridge 1930-32, from the notes of John King and Desmond Lee.* Oxford: Blackwell.

Wittgenstein, Ludwig, 1998. *Culture and Value: A Selection from the Posthumous Remains.* Edited by G. H. von Wright, in collaboration with Heikki Nyman. Text revised by Alois Pichler. Translated by Peter Winch. Oxford: Blackwell.

Wittgenstein, Ludwig, 2003. *Wittgenstein's Nachlass: The Bergen Electronic Edition.* InteLex Past Masters. Charlottesville, Va.: InteLex Corporation.

Wright, Crispin, 1980. *Wittgenstein on the Foundations of Mathematics*. Cambridge, Mass.: Harvard University Press.

Wright, Crispin, 2007. Wang's paradox. In Auxier and Hahn [2007], pp. 415-444.

Bibliography of Hao Wang[1]

Marie Grossi, Montgomery Link, Katalin Makkai, and Charles Parsons

The following is as complete a listing of the published writings of Hao Wang as we have been able to assemble. It is based on a list kept over the years by Marie Grossi in her capacity as Wang's secretary at The Rockefeller University. Some further items, mainly reviews, were uncovered by Katalin Makkai. Some items that we had overlooked, in both English and Chinese, were called to our attention by Hongkuei Kang, to whom we are very grateful.[2]

In the entries for items written in Chinese, Chinese is rendered according to the Pinyin system of romanization.[3] Our information concerning translations of Wang's writings into other languages may be incomplete. Any additions or corrections concerning this or other matters should be sent to Charles Parsons at `parsons2@fas.harvard.edu`.

The order of items within a given year for the most part follows that of Grossi's list. We have, however, placed books first and reviews last. A small number of items are listed with a date some time before their publication because of the long delay in publication; we assume these items circulated in the interim. These are

---

[1]Reprinted with corrections and additions from *Philosophia Mathematica* (3) 6 (1998), 25-38, by permission of the editor, the authors, and Oxford University Press. The bibliography should remain on the *Philosophia Mathematica* web site, so that further updatings can be posted.

[2]Thanks to Richard Jandovitz for his help with this and the earlier published version referred to in note 1. We are indebted to Xing Taotao, Ser-min Shei, Phally Eth, Ma Xiaohe, and Li Dan for supplying further information concerning publications of Wang in Chinese, and to Mihai Ganea, Giovanni Sambin, and the MIT Press for further information concerning translations.

[3]We note that in some earlier items in English, Chinese names occur romanized by the older Wade-Giles system.

indicated by an asterisk on the date.

We wish to thank Robert S. D. Thomas, editor of *Philosophia Mathematica*, for his encouragement of this project.

1944

1944. The metaphysical system of the New Lixue (Chinese). *Zhe xue ping lun. Philosophical Review* 9, no. 3, 39-62.

1945

1945. Language and metaphysics (Chinese). *Zhe xue ping lun. Philosophical Review* 10, no. 1, 35-38. Published in 1946. English translation, [2005].

1947

1947a. Notes on the justification of induction. *The Journal of Philosophy* 44, 701-710. Reprinted in [1974a], Appendix.

1947b. A note on Quine's principles of quantification. *The Journal of Symbolic Logic* 12, 130-132.

1948

1948a. A new theory of element and number. *The Journal of Symbolic Logic* 13, 129-137. Reprinted in [1962a] as chapter XX, section 2.

1948b. The existence of material objects. *Mind* 57, 488-490. Reprinted in [1974a], Appendix.

1948c. Review of Eugene Shen, *Luen li hsueh* (Logic). *The Journal of Symbolic Logic* 13, 215-216.

1949

1949a. A theory of constructive types. *Methodos* 1, 374-384.

1949b. New hopes and old fears. *Chinese Student Opinion* 3, no. 4 (July), 1-3.

1949c. On Zermelo's and von Neumann's axioms for set theory. *Proceedings of the National Academy of Sciences, U. S. A.* 35, 150-155.

1949d. Review of Hermann Weyl, *Philosophy of Mathematics and Natural Science. Physics Today* 2, no. 11 (November), 35-36.

1950

1950a. Remarks on the comparison of axiom systems. *Proceedings of the National Academy of Sciences, U. S. A.* 36, 448-453. Reprinted in [1962a] in chapter XVII.

1950b. The non-finitizability of impredicative principles. *Proceedings of the National Academy of Sciences, U. S. A.* 36, 479-484.

1950c. On scepticism about induction. *Philosophy of Science* 17, 333-335. Reprinted in [1974a], Appendix.

1950d. A proof of independence. *American Mathematical Monthly* 57, 99-100.

1950e. A formal system of logic. *The Journal of Symbolic Logic* 15, 25-32. Reprinted in [1962a] as chapter XVI, section 5.

1950f. Existence of classes and value specification of variables. *The Journal of Symbolic Logic* 15, 103-112. Reprinted in [1962a] in chapter XX, section 1.

1950g. (With J. Barkley Rosser.) Nonstandard models for formal logics. *The Journal of Symbolic Logic* 15, 113-129.

1950h. Set-theoretical basis for real numbers. *The Journal of Symbolic Logic* 15, 241-247. Reprinted in [1962a] as chapter XX, section 3.

1951

1951a. Arithmetic models for formal systems. *Methodos* 3, 217-232.

1951b. Arithmetic translations of axiom systems. *Transactions of the American Mathematical Society* 71, 283-293. Reprinted in [1962a] as chapter XIII, section 4.

1951c. Review of Alfons Borgers, Development of the notion of set and of the axioms for sets. *The Journal of Symbolic Logic* 16, 152-153.

1951d. Review of Paul Lorenzen, Algebraische und logistische Untersuchungen über freie Verbände. *The Journal of Symbolic Logic* 16, 269-272.

1951e. Review of Shen Yu-Ting, Yu-yen, su-hsiang, yu i-i (Language, thought, and meaning). *The Journal of Symbolic Logic* 16, 302-303.

1951f. Review of Yin Fu-Sheng, Characteristics of scientific empiricism and comments thereon (Chinese). *The Journal of Symbolic Logic* 16, 304.

1952

1952a. Truth definitions and consistency proofs. *Transactions of the American Mathematical Society* 73, 243-275. Reprinted in [1962a] as chapter XVIII.

1952b. Negative types. *Mind* 61, 366-368.

1952c. Logic of many-sorted theories. *The Journal of Symbolic Logic* 17, 105-116. Partially incorporated into chapter XII of [1962a] and reprinted in [1990a].

1952d. The irreducibility of impredicative principles. *Mathematische Annalen* 125, 56-66. Reprinted in [1962a] in chapter XVII.

1952e. Review of Arnold Schmidt, Die Zulässigkeit der Behandlung mehrsortiger Theorien mittels der üblichen einsortigen Prädikatenlogik. *The Journal of Symbolic Logic* 17, 76.

1953

1953a. (With Robert McNaughton.) *Les systemes axiomatiques de la théorie des ensembles.* Collection de Logique Mathématique, Série A, no. 4. Paris: Gauthier-Villars. Louvain: E. Nauwelaerts. Russian translation by N. B. Polgrebisski. Moscow, 1963.

1953b. Between number theory and set theory. *Mathematische Annalen* 126, 385-409. Reprinted in [1962a] as chapter XIX.

1953c. Certain predicates defined by induction schemata. *The Journal of Symbolic Logic* 18, 49-59. Reprinted in [1962a] as chapter XXI.

1953d. What is an individual? *Philosophical Review* 62, 413-420. Reprinted in [1974a], Appendix. German translation in W. Stegmüller (ed.), *Das Universalien-Problem*, pp. 280-290. Darmstadt: Wissenschaftliche Buchgesellschaft, 1978.

1953e. The categoricity question of certain grand logics. *Mathematische Zeitschrift* 59, 47-56.

1953f. Quelques notions d'axiomatique. *Revue Philosophique de Louvain* 51, 409-443. French translation of what was later published as chapter I of [1962a].

1953g. A problem on propositional calculus (problem 6). *The Journal of Symbolic Logic* 18, 186.

1953h. Review of G. Kreisel, Note on Arithmetic Models for Consistent Formulae of the Predicate Calculus. *The Journal of Symbolic Logic* 18, 180-181.

1954

1954a. A question on knowledge of knowledge. *Analysis* 14, 142-146. Reprinted in [1974a], Appendix.

1954b. The formalization of mathematics. *The Journal of Symbolic Logic* 19, 241-266. Reprinted in [1962a] as chapter XXIII.

1955

1955a. Undecidable sentences generated by semantic paradoxes. *The Journal of Symbolic Logic* 20, 31-43. Reprinted in [1962a] in chapter XXII.

1955b. Notes on the analytic-synthetic distinction. *Theoria* 21, 158-178. Incorporated into chapter VIII of [1974a]. Chinese translation, *Translations in Philosophy*, no. 1 (1982), 28-38.

1955c. (With G. Kreisel.) Some applications of formalized consistency proofs. *Fundamenta Mathematicae* 42, 101-110. Summarized with part II ([1958c]) in chapter XV, section 2 of [1962a].

1955d. On denumerable bases of formal systems. In Th. Skolem et al., *Mathematical Interpretation of Formal Systems*, pp. 57-84. Amsterdam: North-Holland.

1955e. On formalization. *Mind* 64, 226-238. Reprinted in [1962a] as chapter III and in [1990a]. Also reprinted in Irving M. Copi and James A. Gould (eds.), *Contemporary Readings in Logical Theory.* New York: Macmillan, 1967.

1957

1957a. The axiomatization of arithmetic. *The Journal of Symbolic Logic* 22, 145-158. Reprinted in [1962a] as chapter IV.

1957b. A variant to Turing's theory of computing machines. *Journal of the Association for Computing Machinery* 4, 63-92. Reprinted in [1962a] as chapter VI.

1957c. Universal Turing machines: An exercise in coding. *Zeitschrift für mathematische Logik und Grundlagen der Mathematik* 3, 69-80. Reprinted in [1962a] as chapter VII.

1957d. (With A. W. Burks.) The logic of automata. *Journal of the Association for Computing Machinery* 4, 193-218, 279-297. Reprinted in [1962a] as chapter VIII.

1958

1958a. Eighty years of foundational studies. *Dialectica* 12, 466-497. Also in *Logica: Studia Paul Bernays dedicata*, pp. 262-293. Neuchâtel: Éditions du Griffon, 1959. Reprinted in [1962a] as chapter II.

1958b. Alternative proof of a theorem of Kleene. *The Journal of Symbolic Logic* 23, 250.

1958c. (With G. Kreisel.) Some applications of formalized consistency proofs. Part II. *Fundamenta Mathematicae* 45, 334-335. Summarized with [1955c] in chapter XV, section 2 of [1962a].

1959

1959a. Ordinal numbers and predicative set theory. *Zeitschrift für mathematische Logik und Grundlagen der Mathematik* 5, 216-239. Reprinted in [1962a] as chapter XXV.

1959b. Circuit synthesis by solving sequential Boolean equations. *Zeitschrift für mathematische Logik und Grundlagen der Mathematik* 5, 291-322. Reprinted in [1962a] as chapter X.

1959c. (With G. Kreisel and J. R. Shoenfield.) Number-theoretic concepts and recursive well-orderings. *Archiv für mathematische Logik und Grundlagenforschung* 5, 42-64.

1960

1960a. Toward mechanical mathematics. *IBM Journal of Research and Development* 4, 2-22. Reprinted in [1962a] as chapter IX. Also reprinted in K. Sayre and F. Crosson (eds.), *The Modelling of Mind*, pp. 91-120. University of Notre Dame Press, 1963. (Paperback edition, New York: Simon and Schuster, 1968.) Also reprinted in Jörg Siekmann and Graham Wrightson (eds.), *Automation of Reasoning 1: Classical Papers on Computational Logic, 1957-1966.* Berlin: Springer, 1983. Russian translation in *Problems of Cybernetics.*[4]

---

[4]Grossi's list states that an Italian translation of this paper is in preparation (as of 1995 or earlier). Inquiries in Italy have not yielded a published translation, and it seems likely that none has been published.

1960b. Proving theorems by pattern recognition, Part I. *Communications of the Association for Computing Machinery* 3, 220-234. Reprinted together with Part II ([1961a]), as Bell Technical Monograph 3745. Also reprinted in Siekmann and Wrightson, *op. cit.* (see [1960a]), and in [1990a].

1960c. Symbolic representations of calculating machines. *Summaries of Talks Presented at the Summer Institute for Symbolic Logic, Cornell University, 1957*, pp. 181-188. Princeton, N. J.: Institute for Defense Analyses, Communications Research Division.

1960d. Remarks on constructive ordinals and set theory. *Summaries of talks presented at the Summer Institute for Symbolic Logic, Cornell University, 1957*, pp. 383-390. Princeton, N. J.: Institute for Defense Analyses, Communications Research Division.

1960e. Review of Moh Shaw-Kwei, Simplified introduction to intuitionistic logic (Chinese). *The Journal of Symbolic Logic* 25, 181.

1960f. Review of Moh Shaw-Kwei, About the rules of procedure (Chinese). *The Journal of Symbolic Logic* 25, 182.

1960g. Review of Moh Shaw-Kwei, Axiomatization of many-valued logical systems (Chinese). *The Journal of Symbolic Logic* 25, 181-182.

1960h. Review of Moh Shaw-Kwei, On the explicit form of number-theoretic functions (Chinese). *The Journal of Symbolic Logic* 25, 182.

1960i. Review of Moh Shaw-Kwei, On the definition of primitive recursive functions (Chinese). *The Journal of Symbolic Logic* 25, 182.

1960j. Review of Moh Shaw-Kwei, Some axiom systems for propositional calculus (Chinese). *The Journal of Symbolic Logic* 25, 182-183.

1960k. Review of Moh Shaw-Kwei, On the explicit form of general recursive functions (Chinese). *The Journal of Symbolic Logic* 25, 183.

1961

1961a. Proving theorems by pattern recognition, Part II. *Bell System Technical Journal* 40, 1-41. Also appeared, together with Part I ([1960b]), as Bell Technical Monograph 3745. Reprinted in [1990a].

1961b. Process and existence in mathematics. In Y. Bar-Hillel, E. I. J. Poznanski, M. O. Rabin, and A. Robinson (eds.), *Essays on the Foundations of Mathematics, Dedicated to Prof. A. A. Fraenkel on His 70th Anniversary*, pp. 328-351.[5] Jerusalem: Magnes Press, The Hebrew University of Jerusalem. Partly incorporated into [1974a], chapter VII, and reprinted in [1990a].

1961c. The calculus of partial predicates and its extension to set theory, I. *Zeitschrift für mathematische Logik und Grundlagen der Mathematik* 7, 283-288. Reprinted in [1990a]. Partial Romanian translation by Ilie Parvu in Parvu (ed.), *Epistemologie — Orientari contemporane*. Bucharest: Editura Politica, 1974.

1961d. An unsolvable problem on dominoes. Report BL-30. The Computation Laboratory, Harvard University, 1–5.[6]

1962

1962a. *A Survey of Mathematical Logic*. Peking: Science Press. Also Amsterdam: North-Holland Publishing Company, 1963. Reprinted as *Logic, Computers and Sets*. New York: Chelsea, 1970. Romanian translation by Sorin Vieru and Usher Morgenstern. Bucahrest: Editura Stintififica, 1972.

1962b. (With A. S. Kahr and Edward F. Moore.) Entscheidungsproblem reduced to the $\forall\exists\forall$ case. *Proceedings of the National Academy of Sciences U. S. A.* 48, 365-377.

---

[5]We here follow the title page of the book. Curiously, a reprint of the paper in our possession gives the subtitle as "dedicated to Prof. A. H. Fraenkel on his 70th birthday." 'A. H. Fraenkel' evidently abbreviates the Jewish form of his name, Abraham Halevi Fraenkel.

[6]In [1990a], p. xxiii, Wang states that this paper is included in that book as an appendix to Chapter 9. However, it is not. Hongkuei Kang, who pointed this out to us, states that Wang told him he had forgotten to include it.

1962c. (With Burton Dreben and A. S. Kahr.) Classification of *AEA* formulas by letter atoms. *Bulletin of the American Mathematical Society* 68, 528-532.

1962d. (With A. S. Kahr.) A remark on the reduction problem with application to the AEA formulas (Abstract). *Notices of the American Mathematical Society* 9, 130.

1963

1963a. Mechanical mathematics and inferential analysis. In P. Braffort and D. Hirschberg (eds.), *Computer Programming and Formal Systems*, pp. 1-20. Amsterdam: North-Holland. Reprinted in [1990a].

1963b. (With M. O. Rabin.) Words in the history of a Turing machine with a fixed input. *Journal of the Association for Computing Machinery* 10, 526-527.

1963c. Dominoes and the AEA case of the decision problem. In Jerome Fox et al. (eds.), *Proceedings of the Symposium on the Mathematical Theory of Automata, New York, April 1962*, pp. 23-55. Brooklyn: Polytechnic Press. Reprinted in [1990a].

1963d. Tag systems and lag systems. *Mathematische Annalen* 152, 65-74.

1963e. The mechanization of mathematical arguments. In N. C. Metropolis, A. H. Taub, John Todd, and C. B. Tompkins (eds.), *Experimental Arithmetic, High Speed Computing and Mathematics*, pp. 31-40. Proceedings of Symposia in Applied Mathematics, vol. 15. Providence: American Mathematical Society. Reprinted in [1990a].

1963f. (With A. S. Kahr.) Degrees of RE models of AEA formulas (Abstract). *Notices of the American Mathematical Society* 10, 192–193.

1963g. Review of William and Martha Kneale, *The Development of Logic*. *Mathematical Reviews* 26, 450.

1964

1964a. (With W. V. Quine.) On ordinals. *Bulletin of the American Mathematical Society* 70, 297-298.

1964b. Remarks on machines, sets and the decision problem. In J. N. Crossley and M. A. E. Dummett (eds.), *Formal systems and Recursive Functions*, pp. 304-320. Amsterdam: North-Holland. Reprinted in [1990a].

1964c. Critique [of Robert R. Kofhage, Logic for the computer sciences]. *Communications of the Association for Computing Machinery* 7, 218.

1965

1965a. Russell and his logic. *Ratio* 7, 1-34. Reprinted, revised, as chapter III of [1974a]. German translation in the German edition of *Ratio*. Spanish translation by E. Casaban and E. Garcia, *Teorema*, December 1971, pp. 31-76.

1965b. Formalization and automatic theorem proving. In Wayne A. Kalenich (ed.), *Proceedings of IFIP Congress 65*, pp. 51-58. Washington, D. C.: Spartan Books. Reprinted in [1990a]. Russian translation in *Problems of Cybernetics* 7 (1970), 180-193.

1965c. Note on rules of inference. *Zeitschrift für mathematische Logik und Grundlagen der Mathematik* 11, 193-196.

1965d. Logic and computers. *American Mathematical Monthly* 72, 135-140. Reprinted as chapter IX, section 6 of [1974a] and in [1990a].

1965e. Games, logic and computers. *Scientific American* 213, no. 5 (November), 98-106. Reprinted in [1990a]. Swedish translation in *Modern Datateknik*.

206

1966

1966a. (With S. A. Cook.) Characterizations of ordinal numbers in set theory. *Mathematische Annalen* 164, 1-25.

1966b. (With Kenneth R. Brown.) Finite set theory, number theory and axioms of limitation. *Mathematische Annalen* 164, 26-29.

1966c. (with Kenneth R. Brown.) Short definitions of ordinals. *The Journal of Symbolic Logic* 31, 409-414.

1966d. Russell and philosophy. *The Journal of Philosophy* 63, 670-673. Partly incorporated into in [1974a], chapter XI, section 4.

1967

1967a. Natural hulls and set existence. *Zeitschrift für mathematische Logik und Grundlagen der Mathematik* 13, 175-182. Reprinted in [1990a].

1967b. On axioms of conditional set existence. *Zeitschrift für mathematische Logik und Grundlagen der Mathematik* 13, 183-188. Reprinted in [1990a].

1967c. A theorem on definitions of the Zermelo-von Neumann ordinals. *Zeitschrift für mathematische Logik und Grundlagen der Mathematik* 13, 241-250. Reprinted in [1990a].

1967d. Introductory note to Andrei Nikolaevich Kolmogorov, On the principle of excluded middle. In Jean van Heijenoort (ed.), *From Frege to Gödel, A Source Book in Mathematical Logic, 1879-1931*, pp. 414-416. Cambridge, Mass.: Harvard University Press.

1970

1970a. A survey of Skolem's work in logic. In Th. Skolem, *Selected Works in Logic*, ed. J. E. Fenstad, pp. 17-52. Oslo: Universitetsforlaget. Incorporated, with additions and corrections by J. E. Fenstad, into [2009].

1970b. Remarks on mathematics and computers. In R. B. Banerji and M. D. Mesarovic (eds.), *Theoretical Approaches to Nonnumerical Problem Solving*, pp. 152-160. Berlin: Springer. Partly incorporated into [1974a], chapter IX. Reprinted in [1990a].

1970c. On the long-range prospects of automatic theorem-proving. In M. Laudet, D. Lacombe, L. Nolin, and M. Schützenberger (eds.), *Symposium on Automatic Demonstration*, pp. 101-111. Berlin: Springer. Reprinted in [1990a].

1971

1971a. Logic, computation and philosophy. *L'age de la science* 3, 101-115. Partly incorporated into [1974a], chapter VII. Reprinted in [1990a].

1971b. Letter to the Editor. *The New York Times*, May 30.

1972

1972. Reflections on a visit to China (Chinese). *New China Bimonthly*, no. 7 (October 1), 23-26 and 31. Reprinted in *Xinwan Bao*, November. Second enlarged version, *The Seventies Monthly*, no. 36 (January 1973), 54-60 and no. 37 (February 1973), 85-90; also in *Dagong Bao*, December 1972; and as a separate pamphlet by Bagu Publishing, February 1973. Third revised version, published as a separate pamphlet. Shanghai: The Seventies Publishing Company, 1973. Also in *People's Daily, Reference Information*, March 3, 4, 5, 6, 7, 1973.

1973

1973a. Forty years of culture in Hong Kong (Chinese). *The Seventies Monthly*, no. 44 (September), 22-23.[7]

*1973b. (With Bradford Dunham.) A recipe for Chinese typewriters. IBM report RC4521, September 5. Published in Chinese: see

---

[7]This item is included in Grossi's list, but it appears in *The Seventies Monthly* under the author name Xu Shangwen. See "Hao Wang's Chinese writings" in this volume.

208

[1976a].

1974

1974a. *From Mathematics to Philosophy*. London: Routledge & Kegan Paul. Chapter VI reprinted in Paul Benacerraf and Hilary Putnam (eds.), *Philosophy of Mathematics: Selected Readings*, 2nd ed. Cambridge University Press, 1983. Chapter VII reprinted with slight revisions in Thomas Tymoczko (ed.), *New Directions in the Philosophy of Mathematics*, pp. 131–152. Boston: Birkhäuser, 1986. 2nd ed., Princeton University Press, 1997. Chapter IX, sections 3-4, reprinted (in English) in Christian Thiel (ed.), *Erkenntnistheoretische Grundlagen der Mathematik*, pp. 332-337. Hildesheim: Gerstenberg Verlag, 1982. Italian translation: Torino: Boringhieri, 1984. Chinese translation forthcoming.

1974b. Metalogic. In *Encyclopaedia Britannica*, 15th ed., vol. 11, pp. 1078-1086. Chicago: Encyclopaedia Britannica, Inc. Incorporated into chapter V of [1974a]. Reprinted in part in [1990a].

1974c. Concerning the materialist dialectic. *Philosophy East and West* 24, 303-319.

1974d. Letter to the Editor. *The New York Times*, June 18.

1975

1975a. Notes on a class of tiling problems. *Fundamenta Mathematicae* 82, 295-305. Reprinted in 1990a.

1975b. Letter to the Editor. *Washington Post*, November 1.

1976

1976a. (With Bradford Dunham.) A recipe for Chinese typewriters (Chinese). *Dousou Bimonthly*, no. 14 (March), 56-62. Chinese verson of [*1973b].

1976b. (With Bradford Dunham.) Toward feasible solutions of the tautology problem. *Annals of Mathematical Logic* 10, 117-154.

Reprinted in [1990a].

1976c. Letter to the Editor. *The New York Times*, February 9.

1977

1977a. Large sets. In Robert E. Butts and Jaakko Hintikka (eds.), *Logic, Foundations of Mathematics, Logic and Computability Theory*, pp. 309-333. Dordrecht: Reidel.

1977b. The searchings of Lu Xun (Chinese). *Dousou Bimonthly*, no. 19 (January), 1-14. Abridged version, *The Seventies Monthly*, no. 85 (February), 70-73.

1977c. Dialectics and natural science. *The Overseas Chinese Life Scientists Association Newsletter* 1, no. 2 (March), 48-55.[8]

*1977d. (With D. A. Martin.) Ranked matching and hospital interns. Published in [1990a], pp. 275-289, but dated 1977.

1978

1978. Kurt Gödel's intellectual development. *The Mathematical Intelligencer* 1, no. 3, 182-184. Chinese translation, *Philosophical Problems of Natural Science*, no. 4 (1980), 88-90.

1979

*1979a. Kurt Gödel and some of his philosophical views: On mind, matter, machine and mathematics. *Proceedings of the Roundtable on Aristotle, June 1978*. Paris: UNESCO. Published in French translation; see [1991c].

1979b. Mechanical treatment of Chinese characters (Chinese). *Dianzi Jisuanji Dongtai*, no. 6, 1-4. English translation, On information processing in the Chinese language, by Fan Lanying of the Beijing Institute of Computer Technology in [1990a].

---

[8]We have not been able to locate this item or determine with certainty whether it is written in English or Chinese.

210

1979c. China today and its development over the last sixty years (Chinese). *Wide Angle Monthly*, no. 86, 32-49.

1980

1980. Kurt Gödel. In *McGraw-Hill Encyclopedia of Scientists and Engineers*, pp. 438-439. New York: McGraw-Hill.

1981

1981a. *Popular Lectures on Mathematical Logic*. Beijing: Science Press. Also New York: Van Nostrand Reinhold. Reprinted with a Postscript, New York: Dover Publications, 1993. Chinese translation: Beijing: Science Press, 1981.

1981b. Remeeting Mr. Shen Congwen (Chinese). *Hai Nei Wei*, 28, 25-26. Reprinted in *Dadi*, no. 2, 27-28.

1981c. Some facts about Kurt Gödel. *The Journal of Symbolic Logic* 46, 653-659. Reprinted with some revisions in [1987a], chapter 2.

1981d. Specker's mathematical work from 1949 to 1979. *L'Enseignement mathématique* 72, 85-98. Also in Erwin Engeler, Hans Läuchli, and Volker Strassen (eds.), *Logic and Algorithmic: An International Symposium Held in Honour of Ernst Specker, Zürich, February 5-11, 1980*, pp. 11-24. Monographie de *L'enseignement mathématique* no. 30. Genève: L'enseignement mathématique, Université de Genève.

1981e. Gödel and Wittgenstein (Chinese). *Philosophical Research Monthly*, no. 3, 25-37.

1981f. Mathematical logic (Chinese). *Problems of Natural Sciences*, no. 3, 70-71.

1982

1982a. To confirm some impressions by Lu Xun (Chinese). *Dushu Monthly*, April, 70-76.

1982b. Memories related to Professor Jin Yuelin (Chinese). *Wide Angle Monthly*, no. 122, 61-63. Reprinted in *Chinese Philosophy* 11 (1984), 487–493. Also reprinted in Liu Peiyu (ed.), *The Reminiscences of Jin Yuelin and Reminiscences about Jin*, pp. 161-167. Chengdu: Sichuan Educational Press, 1995. English translation in this volume; see [2011a].

1983

1983. Philosophy: Chinese and Western. *Commentary: Journal of the National University of Singapore Society* 6, no. 1 (September), 1-9.

1984

1984a. The formal and the intuitive in the biological sciences. *Perspectives in Biology and Medicine* 27, 525-542.

1984b. Computer theorem proving and artificial intelligence. *Contemporary Mathematics* 29, 49-70. Reprinted in [1990a].

1984c. Wittgenstein's and other mathematical philosophies. *The Monist* 67, 18-28.

1984d. Thought and action. *South China Morning Post, The Hong Kong Standard*, June 1.

1985

1985a. *Beyond Analytic Philosophy. Doing Justice to What We Know*. Cambridge, Mass.: MIT Press. Paperback edition, 1987.

1985b. Two commandments of analytic empiricism. *The Journal of Philosophy* 82, 449-462. Partly incorporated into [1985a], Introduction. Chinese translation, *Social Sciences in China*, no. 4 (July 1985).

1986

1986a. China and Western philosophy (Chinese). *Chinese Culture Quarterly* 1, no. 1 (September), 39-60.

1986b. Quine's logical ideas in historical perspective. In Lewis Edwin Hahn and Paul Arthur Schilpp (eds.), *The Philosophy of W. V. Quine*, pp. 623-643. La Salle, Ill.: Open Court. 2nd ed., 1998.

1986c. Gödel's and some other examples of problem transmutation (Chinese). *Zhe xue yi cong. Collection of Translations of Philosophy* 6, 62-66.[9]

1987

1987a. *Reflections on Kurt Gödel.* Cambridge, Mass.: The MIT Press. Paperback edition, 1990. Japanese translation: Sangyo Tosho, 1988. French translation: Paris: Armand Colin, 1991. Spanish translation: Alianza, 1992. Korean translation: Seoul: Minumsa Publishing, 1997. Chinese translation, with a new preface: Shanghai: Shanghai Translation Publishing House, 1997.

1987b. Gödel and Wittgenstein. In Paul Weingartner and Gerhard Schurz (eds.), *Logic, Philosophy of Science and Epistemology*, pp. 83-90. Proceedings of the 11th International Wittgenstein Symposium, Kirchberg am Wechsel, Austria, 4-13 August, 1986. Vienna: Verlag Hölder-Pichler-Tempsky.

1987c. On distinguishing problems of different orders (Chinese). *Chinese Culture Quarterly* 1, no. 4 (summer), 35-40.

1987d. Einstein and Gödel: Contrast and friendship (Chinese). *Journal of Tsinghua University* 2, no. 1, 32-39, 56. Revised and enlarged version in Chinese, Exploring the eternal: Gödel and Einstein. *Twenty-First Century Bimonthly*, no. 2 (December 1990), 72–81.

---

[9]This journal has been known as the *Shi jie zhe xue* since 2002. Grossi's list identifies the journal where this paper appeared as *Dialectics of Nature*.

1987e. The way of Jin Yuelin (Chinese). In Institute of Philosophical Research, Chinese Academy of Social Science (ed.), *Studies in Jin Yuelin's Thought*, pp. 45-50. Chengdu: Sichuan People's Publishing.

1987f. Review of Galvano Della Volpe, *Logic as a Positive Science* (Chinese). *Chinese Culture Quarterly* 1, no. 3 (spring), 101-104. Translated from English by Ser-min Shei.

1987g. Review of David Rubinstein, *Marx and Wittgenstein: Social Praxis and Social Explanation* (Chinese). *Chinese Culture Quarterly* 1, no. 3 (spring), 104-107. Translated from English by Ser-min Shei.[10]

1989

1989a. Tharp and conceptual logic. *Synthese* 81, 141-152.

1989b. A reading of Wang You-qin on Lu Xun (Chinese). *W. M. Semi-Annual*, no. 2 (July), 118-133.

1990

1990a. *Computation, Logic, Philosophy. A Collection of Essays.* Beijing: Science Press. Dordrecht: Kluwer Academic Publishers.

1990b. Philosophy through mathematics and logic. In Rudolf Haller and Johannes Brandl (eds.), *Wittgenstein–Towards a Re-evaluation*, pp. 142-154. Proceedings of the 14th International Wittgenstein Symposium, Centenary Celebration, Kirchberg am Wechsel, Austria, 1989. Vienna: Verlag Hölder-Pichler-Tempsky.

1990c. Mind, brain, machine. *Jahrbuch 1990 der Kurt-Gödel-Gessellschaft*, 5-43. Proceedings of the First Kurt Gödel Colloquium, Salzburg, Austria, September 1989.

1990d. Aperiodicity and constraints. *Jahrbuch 1990 der Kurt-Gödel-Gessellschaft*, 88-93.

---

[10]Shei, who told of these items, states that he no longer has the English originals. However, copies may be in the Hao Wang papers at the Rockefeller University Archives.

214

1990e. Between philosophy and literature (Chinese). *Dushu Monthly* (April), 58–66.

1991

1991a. Gödel's and some other examples of problem transmutation. In T. Drucker (ed.), *Perspectives on the History of Mathematical Logic*, pp. 101-109. Boston: Birkhäuser. Chinese translation, [1986c].

1991b. To and from philosophy — Discussions with Gödel and Wittgenstein. *Synthese* 88, 229-277.

1991c. Kurt Gödel et certaines de ses conceptions philosophiques: l'esprit, la matière, la machine et les mathématiques. In G. Hahn and M. A. Sinaceur (eds.), *Penser avec Aristote. Etudes réunies sous la direction de M. A. Sinaceur*, pp. 441-451. Toulouse: Erès. (Cf. *1979a.)

1991d. Gödel and Einstein as companions. In John Brockman (ed.), *Doing Science. The Reality Club*, pp. 282-294. New York: Prentice Hall Press.

1993

1993a. Imagined discussions with Gödel and with Wittgenstein. *Jahrbuch 1992 der Kurt-Gödel-Gesellschaft*, pp. 3-49.

1993b. Can bodies or computers have souls? I. Psychophysical parallelism and algorithmism for the physical world (Chinese). *Twenty-First Century Bimonthly*, no. 15 (February), 102-110.

1993c. Can bodies or computers have souls? II. On algorithmism of the mind and the problem of feasibility (Chinese). *Twenty-First Century Bimonthly*, no. 16 (April), pp. 72-78.

1993d. On physicalism and algorithmism: Can machines think? *Philosophia Mathematica* (III) 1, 97-138.

1993e. What is logic? In Klaus Puhl (ed.), *Wittgenstein's Philos-*

ophy of Mathematics, pp. 11-23. Proceedings of the 15th International Wittgenstein Symposium, part 2. Vienna: Verlag Hölder-Pichler-Tempsky. Also *The Monist* 77 (1994), 261-274.

1993f. From Kunming to New York (Chinese). *Dushu Monthly* (May), 140–143. English translation in this volume; see [2011b].

1993g. New directions in science and in society: From traditions to innovations. In *Nineteenth World Congress of Philosophy, Moscow, August 1993, Book of Abstracts: Invited Lectures*, pp. 52–59. Moscow.

1995

1995. Time in philosophy and in physics: From Kant and Einstein to Gödel. *Synthese* 102, 215-234.

1996

1996a. *A Logical Journey. From Gödel to Philosophy*. Cambridge, Mass.: The MIT Press.

1996b. Skolem and Gödel. *Nordic Journal of Philosophical Logic* 1, 119-132.

2005

2005. Language and metaphysics. Translation by Richard Jandovitz and Montgomery Link of [1945]. *Journal of Chinese Philosophy* 32, no. 1, 139-147.

2009

2009. (With Jens-Erik Fenstad.) Thoralf Albert Skolem. In Dov Gabbay and John Woods (eds.), *Handbook of the History of Logic*, volume 5: *Logic from Russell to Church*, pp. 127-194. Amsterdam: North-Holland.

2011

2011a. Memories related to Professor Jin Yuelin. Translation by Montgomery Link and Richard Jandovitz of [1982b]. In this volume.

2011b. From Kunming to New York. Translation by Richard Jandovitz and Montgomery Link of [1993f]. In this volume.

2011c. Sets and concepts, on the basis of discussions with Gödel. Edited with an introduction by Charles Parsons. In this volume.